CHAIN STRUCTURE
AND CONFORMATION
OF MACROMOLECULES

CHAIN STRUCTURE
AND CONFORMATION
OF MACROMOLECULES

Frank A. Bovey
Bell Laboratories
Murray Hill, New Jersey

with a chapter by
Lynn W. Jelinski
Bell Laboratories
Murray Hill, New Jersey

ACADEMIC PRESS 1982
A Subsidiary of Harcourt Brace Jovanovich, Publishers

New York   London
Paris   San Diego   San Francisco   São Paulo   Sydney   Tokyo   Toronto

ACADEMIC PRESS, INC.
111 Fifth Avenue, New York, New York 10003

United Kingdom Edition published by
ACADEMIC PRESS, INC. (LONDON) LTD.
24/28 Oval Road, London NW1 7DX

Library of Congress Cataloging in Publication Data

Bovey, Frank Alden, Date.
  Chain structure and conformation of macromolecules.

  Includes index.
  1. Macromolecules.   I. Title.
QD381.B657   1982   547.7   82-20779
ISBN 0-12-119780-8

PRINTED IN THE UNITED STATES OF AMERICA

82 83 84 85   9 8 7 6 5 4 3 2 1

CONTENTS

CHAPTER 8 **Solid State NMR of Macromolecules**

L. W. JELINSKI

PREFACE

This book represents a course of lectures delivered in 1981 in the Department of Chemical Engineering at the Massachusetts Institute of Technology. Its principal purpose was to provide an introduction to the chain structures of synthetic polymers and their determination in solution and in the solid state. The treatment is at the undergraduate or first-year graduate level and is primarily fundamental rather than technological in orientation. We shall not discuss synthetic methods and polymerization mechanisms except as they are related very directly to details of chain structure, such as comonomer sequences in copolymers.

Chapter 1 will be an introduction to the structure of polymer chains, including a brief history of the macromolecular concept and of stereochemical and geometrical isomerism in synthetic polymer chains. Chapter 2 introduces vibrational spectroscopy and nuclear magnetic resonance spectroscopy; the latter will receive particularly heavy emphasis because of its great power in polymer structure determination. Chapter 3 deals in detail with stereochemical configuration, while Chapter 4 treats geometrical isomerism in diene copolymers. Chapter 5 describes copolymerization and the measurement of copolymer structure. Chapter 6 deals with "regioregularity," i.e., head-to-tail versus head-to-head:tail-to-tail isomerism and also the important subject of branching. Chapter 7 describes at some length polymer chain conformation, including a discussion of the rotational isomeric state method of calculation of polymer chain dimensions.

Chapter 8, by Dr. Lynn W. Jelinski, provides an unusual feature of the book, as it deals with the NMR observation of polymers in the solid state by the method of "magic angle" spinning, by which both high resolution structural information and dynamic measurements are possible.

I thank the members of the staff of the Chemical Engineering Department of the Massachusetts Institute of Technology, particularly Professor E. W. Merrill, for their encouragement in the preparation of this volume. I particularly thank Dr. Lynn Jelinski for her invaluable and indefatigable efforts not only in providing Chapter 8 but, in general, helping in organizing and supervising the preparation of the manuscript and the production of the book.

F. A. BOVEY

Chapter 1

THE STRUCTURE
OF POLYMER CHAINS

1.1 INTRODUCTION

In this relatively brief discussion, we shall cover a number of important areas of polymer science. We shall consider a variety of topics embraced under the heading of polymer chain *microstructure*. What does this term mean? We shall deal at some length with *stereochemical configuration*. Why is it significant and how can it be observed? We shall discuss *regiospecificity* and its importance and measurement. The formation and architecture of *copolymers* will require our attention. Polymer chain *conformation*, i.e., those forms of chain isomerism that result from rotation about bonds without alteration of the covalent structure, will be discussed in detail.

We shall not discuss synthesis and polymerization mechanisms in general, but only when it may aid our understanding of structure; in connection with copolymer structure and chain branching, this will involve detailed, if rather specialized, consideration of free radical mechanisms.

In treating some of these topics, there will be a fairly heavy emphasis on *nuclear magnetic resonance (NMR) spectroscopy*: first, because it is an extraordinarily powerful technique for the study of polymers; and second, because it happens to be the authors' field of specialization. This may at times lead to an impression of imbalance, but we hope to be able to convince the reader that NMR deserves the attention it will receive. Vibrational spectroscopy will also be described and certain applications considered, but its treatment will be much more limited.

1.2 HISTORY OF THE MACROMOLECULAR CONCEPT

In order to discuss intelligently the structures of polymer chains, we must first agree that polymers are indeed composed of long-chain molecules or *macromolecules*. Even at a time—the period of approximately 1900 to 1930—when organic chemistry was in a fairly advanced state of development, the idea that there could be covalently bonded molecules with molecular weights as high as 100,000 or even greater was not accepted. It was evidently felt that such structures would be inherently unstable. Another influence was the notion, strongly held by organic chemists, that every substance whatever—including materials we now recognize as macromolecular—had to be describable by a definite molecular formula. Indeed, a very impressive body of knowledge of small, crystallizable compounds had been built up from known synthetic and degradative reactions, supplemented by elemental analysis and physical techniques, principally mixed melting points. The idea of long chains was bad enough; the idea of a *mixture* of chains of varying length and perhaps varying stereochemistry as well, neither clearly specified, was even more repugnant.

In the early days of this century it was commonly assumed that polymers were aggregates of small molecules held together by vaguely defined "secondary" or "partial" valence forces. Thus, Harries (1904, 1905) proposed that natural rubber was composed of isoprene units combined as dimers (as indicated by ozonization to levulinic aldehyde, $CH_3COCH_2CH_2CHO$), which were then aggregated as shown. These secondary valence forces were thought to require the presence of double bonds, and Staudinger (Staudinger and Fritschi, 1922) was able to shake severely the foundations of the association theory by showing that on hydrogenation to a saturated hydrocarbon natural rubber retained its polymeric character, having still no sharp melting point and giving solutions of high viscosity.

The secondary valence forces, although undefined, had to be regarded as strong, as the cyclic molecule could not be distilled from the supposed aggregate, although it was well known that isoprene itself could be recovered when natural rubber is pyrolyzed, as observed in 1860 by Greville Williams (Williams, 1860).

Gradually the macromolecular hypothesis overcame all opposition. Three principal reasons for this have been cited by Mark (1967):

(i)   The observation of Katz (1925) that stretched natural rubber shows a crystalline X-ray fiber diagram, and the related observation by Meyer and Mark (1928) of crystallinity in cellulose, were best explained by long-chain molecules forming a semicrystalline structure.

(ii)   Solution viscosity measurements (Staudinger, 1928) and ultracentrifugation observations (Svedberg, 1926) were best explained on the basis of large molecules. It was also observed that pure preparations of proteins showed very narrow molecular weight distributions, a finding particularly difficult to explain on the basis of aggregations of small molecules.

(iii)   Finally, and probably most convincingly, Carothers (1931) and his colleagues at the duPont company were able to secure the argument for the macromolecular hypothesis by carrying out straightforward polycondensation reactions that could lead only to long-chain molecules.

These matters are discussed by Flory in the first chapter of his classic book (Flory, 1953) to which the reader is referred for a very full and excellent treatment.

1.3 ISOMERISM IN VINYL POLYMER CHAINS

Although there is a vast variety of synthetic polymers, we shall be mainly concerned with vinyl and diene polymer (and copolymer) chains, since these present the most intriguing structural problems.

Let us consider the types of isomerism that may be implied in the deceptively simple reaction:

$$n \quad \begin{matrix} A \\ \ \\ B \end{matrix} C=C \begin{matrix} H \\ \ \\ H \end{matrix} \longrightarrow \begin{bmatrix} A & H \\ C-C \\ B & H \end{bmatrix}_n$$

1.3.1 Head to Tail (i) Versus Head to Head:Tail to Tail (ii) Isomerism

(i) ... $-\overset{\underset{B}{A}}{C}-\overset{\underset{H}{H}}{C}-\overset{\underset{B}{A}}{C}-\overset{\underset{H}{H}}{C}-\overset{\underset{B}{A}}{C}-\overset{\underset{H}{H}}{C}-\overset{\underset{B}{A}}{C}-\overset{\underset{H}{H}}{C}-$ ...

(ii) ... $-\overset{\underset{B}{A}}{C}-\overset{\underset{H}{H}}{C}-\underbrace{\overset{\underset{H}{H}}{C}-\overset{\underset{B}{A}}{C}}-\overset{\underset{B}{A}}{C}-\overset{\underset{H}{H}}{C}-\overset{\underset{B}{A}}{C}-\overset{\underset{H}{H}}{C}-$ ...

inverted unit

A head-to-head junction with no accompanying tail-to-tail unit will also arise from the recombination of growing chain radicals; in this case, of course, there can be only one such unit per chain:

$$\cdots-CH_2-\overset{\underset{B}{A}}{C}-CH_2-\overset{\underset{B}{A}}{C}\cdot \ + \ \cdot\overset{\underset{B}{A}}{C}-CH_2-\overset{\underset{B}{A}}{C}-CH_2-\cdots$$

$$\downarrow$$

$$\cdots-CH_2-\overset{\underset{B}{A}}{C}-CH_2-\overset{\underset{B}{A}}{C}-\overset{\underset{B}{A}}{C}-CH_2-\overset{\underset{B}{A}}{C}-CH_2-\cdots$$

1.3.2 Stereochemical Configuration

This term refers to the *relative handedness* of successive monomer units. In vinyl polymers the main-chain substituted carbons, commonly designated as α-carbons, are termed "pseudoasymmetric" since, if the chain ends are disregarded, such carbons do not have the four different substituents necessary to qualify for being truly asymmetric. Nevertheless, they have the possibility of relative handedness. The simplest regular arrangements along a chain are the *isotactic* structure (Fig. 1.1a), in which all the substituents, here represented by *R*, are located on the same side of the zigzag plane representing the chain stretched out in an all-trans conformation (conformational matters are dealt with in Chapter 7); and the *syndiotactic* arrangement, in which the groups alternate from side to side (Fig. 1.1b). In the *atactic* arrangement, the *R* groups are placed at

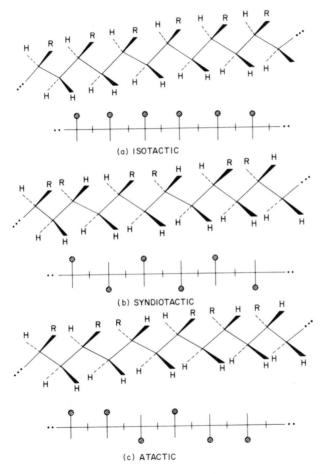

(a) ISOTACTIC

(b) SYNDIOTACTIC

(c) ATACTIC

Fig. 1.1.   Schematic representation of (a) isotactic, (b) syndiotactic, and (c) atactic chains.

random on either side of the zigzag plane (Fig. 1.1c). It should be emphasized that these isomeric forms cannot be interconverted by rotating *R* groups about the carbon—carbon bonds of the main chain.

The possibility of such isomerism had been pointed out by Staudinger (1932), Huggins (1944), and Schildknecht *et al.* (1948). But it was not taken very seriously because there seemed to be no effective way to control it, nor was it believed to have any marked effect on polymer properties. Interest became intense, however, when Natta (1955) showed that by use of the coordination catalysts developed by Ziegler for the polymerization of ethylene, α-olefins can be polymerized to products having stereoregular structures. A major revolution in polymer science and technology was thus initiated, which continues to this day.

1,2-Disubstituted vinyl monomers have additional options in that the polymer chain now has two interleaved systems of nonidentical

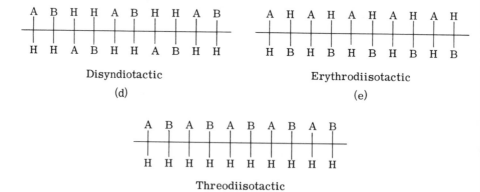

pseudoasymmetric carbon atoms. Three limiting configurations can be defined: *disyndiotactic* (d), *erythrodiisotactic* (e), and *threodiisotactic* (f):

The prefixes erytho- and threo- are based on the configurations of the sugars erythrose and threose. A mnemonic device that may be useful is to recall that if A=B, then erythro would be meso and threo would be racemic:

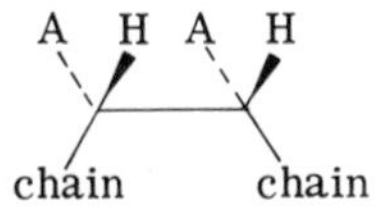

A H A H   A H H A

chain   chain   chain   chain

meso 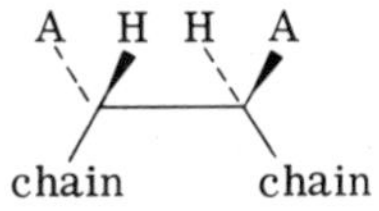   *racemic*

There are a very limited number of actual cases in which A and B are equivalent, and we shall consider these later.

Vinyl polymers may acquire true asymmetric centers by ring closure. The ring may be closed between α-carbons, as in poly(vinyl formal):

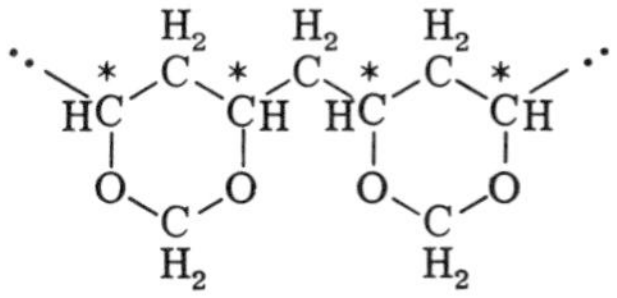

$$
\begin{array}{c}
\ \ \ \ \ H_2 \ \ \ \ H_2 \ \ \ \ H_2 \\
\cdot \cdot \diagdown \ \overset{*}{C} \diagup \overset{*}{C} \diagdown \overset{*}{C} \diagup \cdot \cdot \\
HC \diagup \ \diagdown CH \ \ HC \diagdown \ \diagup CH \\
O \diagdown \ \diagup O \ \ \ O \diagdown \ \diagup O \\
\ \ C \ \ \ \ \ \ \ C \\
\ \ H_2 \ \ \ \ \ \ H_2
\end{array}
$$

Ring formation may occur between adjacent carbons; some of the possibilities are as follows, the rings being represented in a generalized manner:

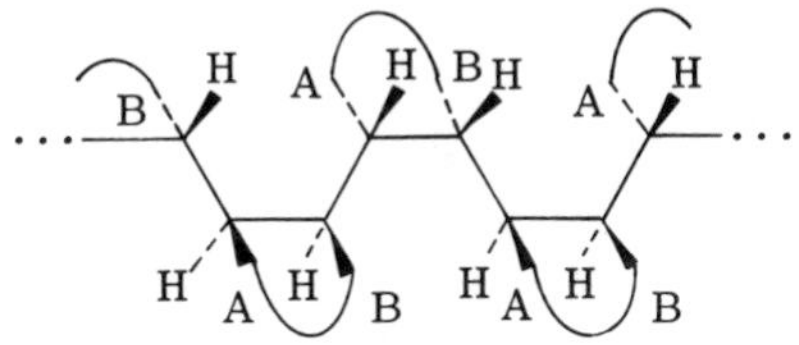

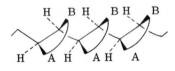

Erythrodiisotactic (ring is *cis*)

(g)

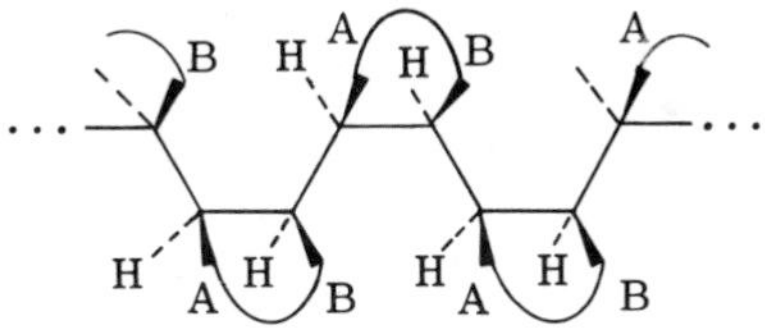

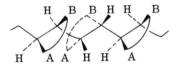

Erythrodisyndiotactic (ring is *cis*)

(h)

H B H   B H   B

H   A H   A H   A

Threodiisotactic (ring is *trans*)

(i)

H   B B H H   B

H   A A H H   A

Threodisyndiotactic (ring is *trans*)

(j)

The rings may be symmetrical (A=B) or unsymmetrical (A≠B). An example of a symmetrical ring system is polycyclobutene:

cis   *trans*

It may be noted that the cis unit has a plane of symmetry and is meso, whereas the trans unit is chiral and racemic, i.e. it may be *d* or *l* (alternatively *R* or *S*).

Polybenzofuran is an example of an unsymmetrical ring system:

1.4 ISOMERISM IN DIENE POLYMER CHAINS

The polymerization of diene monomers can produce structures having combinations of geometrical and stereochemical isomerism. Butadiene can give 1,4 or 1,2 enchainment. The 1,4 structures may be (a) cis (Z) or (b) trans (E):

$$
\begin{array}{cc}
\underset{\cdots\,-CH_2\quad CH_2-\cdots}{\overset{H\qquad H}{C=C}} & \underset{\cdots\,-CH_2\quad H}{\overset{H\qquad CH_2-\cdots}{C=C}} \\[2mm]
1,4\text{-}cis\ (Z) & 1,4\text{-}trans\ (E) \\
\text{(a)} & \text{(b)}
\end{array}
$$

The 1,2 structures have the same configurational properties as vinyl polymers and may occur in (c) isotactic or (d) syndiotactic sequences:

$$
\begin{array}{cc}
\cdots\,-CH_2-\underset{\underset{CH_2}{\overset{\|}{CH}}}{\overset{}{CH}}-CH_2-\underset{\underset{CH_2}{\overset{\|}{CH}}}{\overset{}{CH}}-\cdots &
\cdots\,-CH_2-\underset{\underset{CH_2}{\overset{\|}{CH}}}{\overset{}{CH}}-CH_2-\underset{\underset{CH}{\overset{\|}{CH_2}}}{\overset{}{CH}}-\cdots \\[3mm]
1,2\text{-Isotactic} & 1,2\text{-Syndiotactic} \\
\text{(c)} & \text{(d)}
\end{array}
$$

Polymers of 2-substituted butadienes, an important type, may show every sort of isomerism we have discussed so far. 1,4 Units may exhibit (e) head-to-tail and (f) head-to-head:tail-to-tail structures:

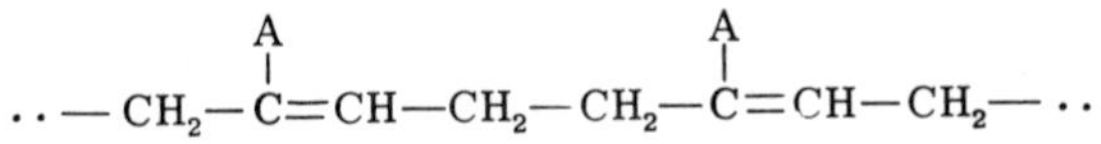

$$\cdots -CH_2-\overset{\overset{\textstyle A}{|}}{C}=CH-CH_2-CH_2-\overset{\overset{\textstyle A}{\uparrow}}{C}=CH-CH_2-\cdots$$

Head-to-tail

(e)

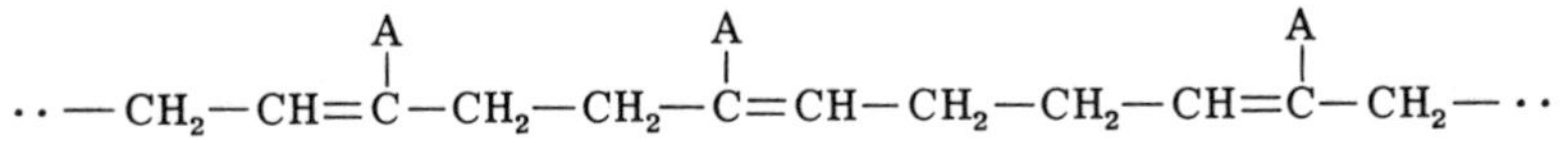

$$\cdots -CH_2-CH=\overset{\overset{\textstyle A}{|}}{C}-CH_2-CH_2-\overset{\overset{\textstyle A}{|}}{C}=CH-CH_2-CH_2-CH=\overset{\overset{\textstyle A}{|}}{C}-CH_2-\cdots$$

Head-to-head : tail-to-tail

(f)

The proportion of inverted units may be substantial, for example, when A=Cl (polychloroprene or Neoprene), which we shall discuss in Section 4.4.

In addition to 1,4 addition, either in a (g) cis or (h) trans fashion,

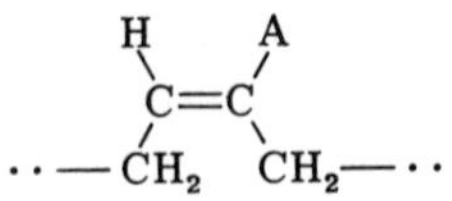

1,4-*cis* (Z)

(g)

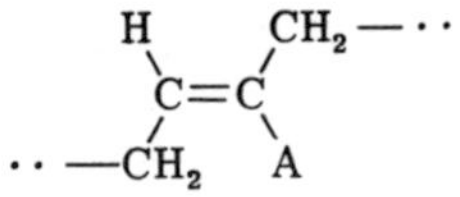

1,4-*trans* (E)

(h)

such monomers have a choice between (i) 1,2 or (j) 3,4 addition, both of which may occur in isotactic or syndiotactic sequences (and head-to-tail and head-to-head:tail-to-tail):

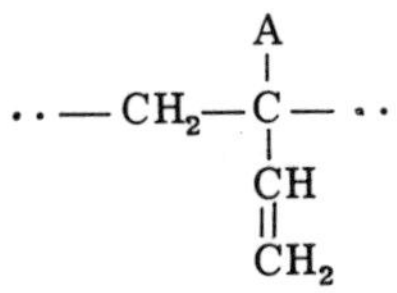

1,2

(i)

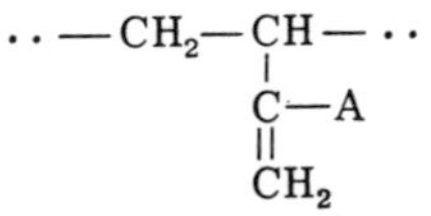

3,4

(j)

When $A=CH_3$ we have, of course, polyisoprene. In nature, this occurs as the cis isomer—natural rubber—and as the trans isomer, called *balata* or *gutta percha*. Each is very pure, without a trace of the other isomer. (They are formed by complex, enzyme-catalyzed processes not involving isoprene as such.) Balata is a semicrystalline plastic rather than a rubber. *Gutta percha* comes from a different plant but is no longer produced commercially. Both isomeric chains can also be prepared synthetically, but are then not entirely stereochemically pure. This will be discussed further in Chapter 4.

The 1-substituted 1,3-butadienes such as 1,3-pentadiene can yield all of the types of isomers we have just discussed and still others in addition. Thus, Natta *et al.* (1963) prepared an optically active *cis*-1,4-poly-1,3-pentadiene; this is possible with a suitably chosen optically active catalyst (see next section) because the chain now has true asymmetric centers:

Such monomers can yield cis-isotactic, trans-isotactic, cis-syndiotactic, and trans-syndiotactic structures, but as yet these have not been prepared and established.

Dienes of the $ACH=CHCH=CHA$ type can give rise to the following isomeric chains by 1,4 addition:

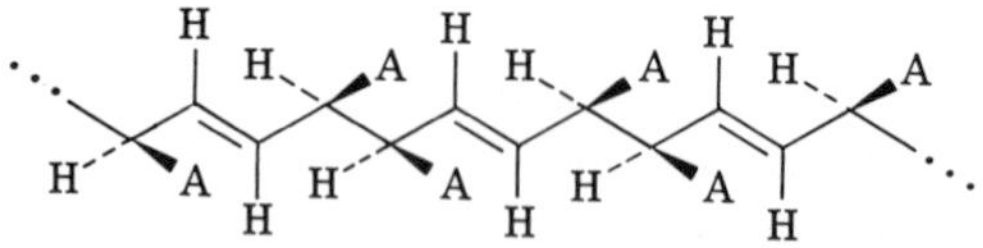

trans-dl

(k)

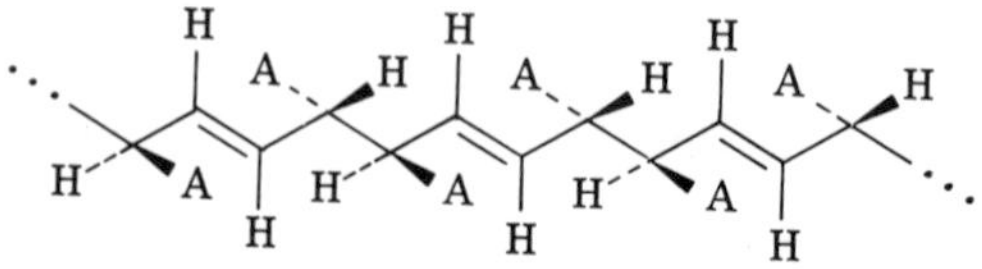

trans-meso

(l)

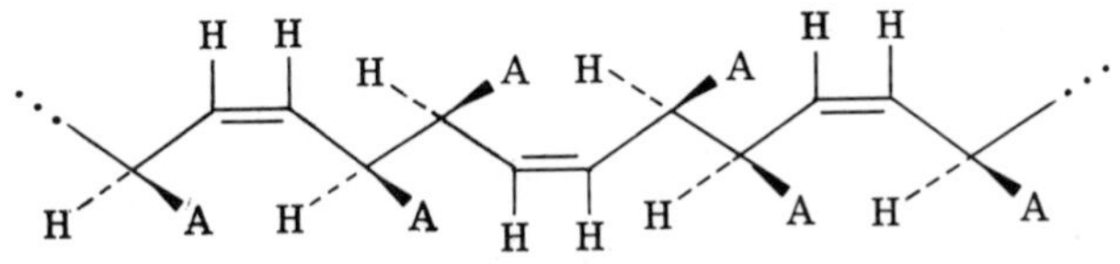

cis-dl

(m)

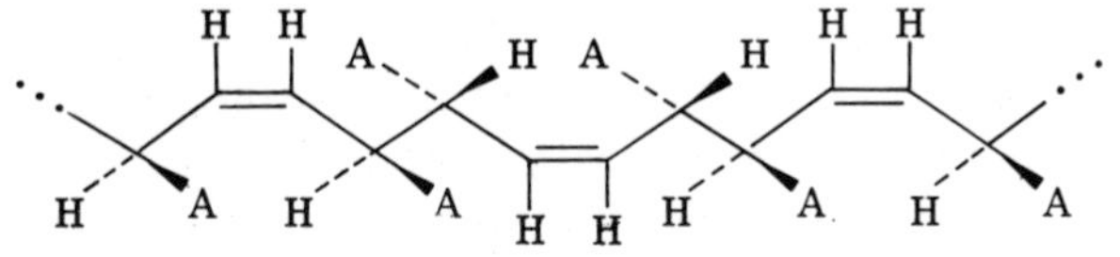

cis-meso

(n)

If the substituents are unlike, ACH=CHCH=CHB, the following structures can be generated:

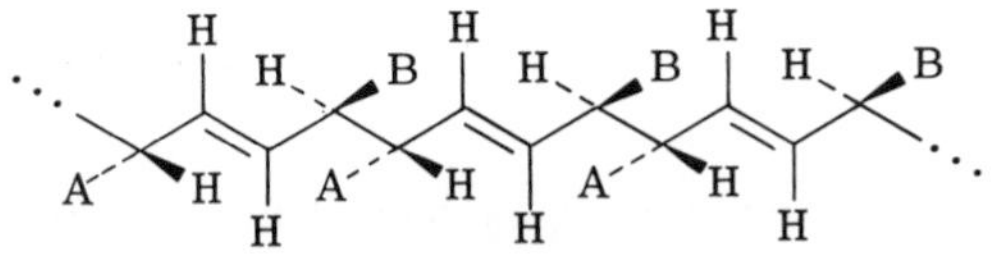

dl-trans-erythro

(o)

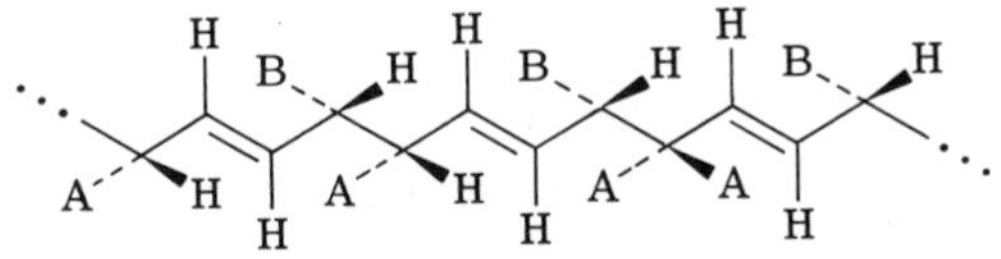

dl-trans-threo

(p)

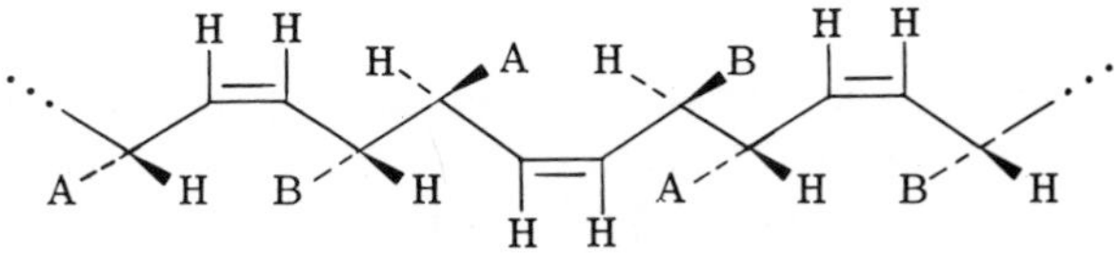

dl-cis-erythro

(q)

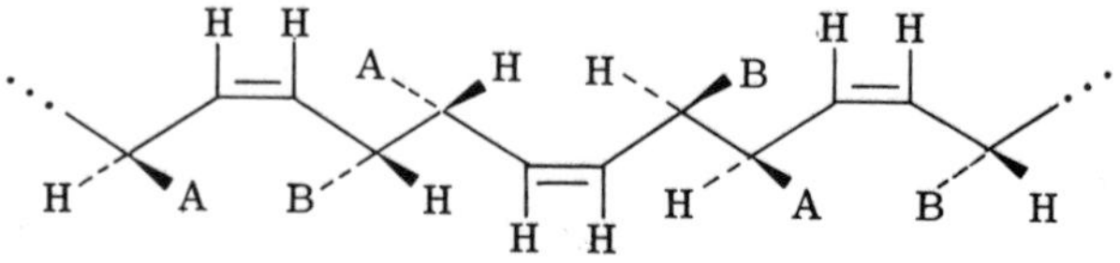

$$dl\text{-}cis\text{-}threo$$

(r)

These structures may be termed diisotactic; one can also imagine syndiotactic structures with A or A and B alternating from side to side of the plane of the main-chain carbons. Natta *et al.* (1960; Dall'Asta *et al.*, 1962) have prepared optically active crystalline polymers from esters of *trans-trans* sorbic acid:

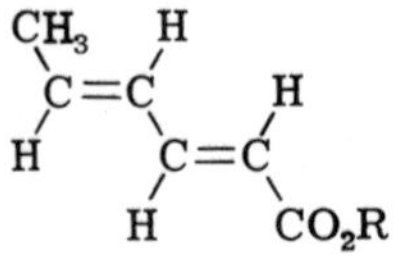

1.5 VINYL POLYMERS WITH OPTICALLY ACTIVE SIDE CHAINS

Vinyl and related monomers may have true asymmetric centers in the side-chains, for example,

$$RCH_2\overset{\overset{\displaystyle R'}{\underset{*}{|}}}{C}H(CH_2)_n CH{=}CH_2$$

$$RCH_2\overset{\overset{\displaystyle R'}{\underset{*}{|}}}{C}H(CH_2)_n OCH{=}CH_2$$

$$RCH_2\overset{\overset{\displaystyle R'}{\underset{*}{|}}}{C}H(CH_2)_n \overset{\overset{\displaystyle H}{|}}{C}{=}O$$

(It may be noted that aldehydes can be polymerized through the carbonyl double bond and thus in effect behave like vinyl monomers.) These asymmetric centers will be carried over into the polymers. Polymerization of the optically active enantiomer (*d* or *l*; or *R* and *S*) will lead to an optically active polymer that may also be isotactic, syndiotactic, or atactic in the usual sense.

Polymerization of the racemic (*RS*) monomer will generally result in a copolymer (see Chapter 5) of the *R* and *S* monomer units, both coexisting in the same chain. However, it has been observed that some *RS* monomers, for example *dl*-4-methyl-1-hexene, can be polymerized by nonchiral catalysts to yield isotactic polymers which can be separated into poly (*R*) and poly (*S*) chains, a very surprising and intriguing observation (Pino *et al.*, 1963):

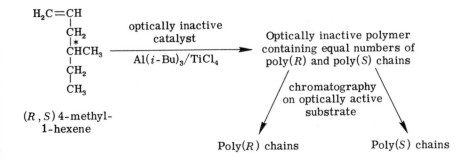

It is evident that under the influence of this type of catalyst a growing chain ending in an *R* unit can only add another *R* unit, while a growing chain ending in an *S* unit can only add another *S* unit. This type of process has been termed *stereoselective polymerization*.

There is another type of selectivity which is exerted by a *chiral* catalyst. The conditions must be right. For example, if we use the chiral catalyst

$$\left(\begin{array}{c} \text{CH}_3 \\ \backslash \overset{*}{\text{C}}\text{HCH}_2 \\ / \\ \text{CH}_3\text{CH}_2 \end{array} \right)_3 \text{Al}/\text{TiCl}_4$$

with the same monomer, (*R,S*) 4-methyl-1-hexene, we obtain a result similar to that observed with the nonchiral catalyst. But if the asymmetric center is one carbon closer to the double bond, as in 3-methyl-1-pentene, the catalyst will now select one enantiomer (not necessarily the one with the same handedness as the catalyst) and will ignore the other:

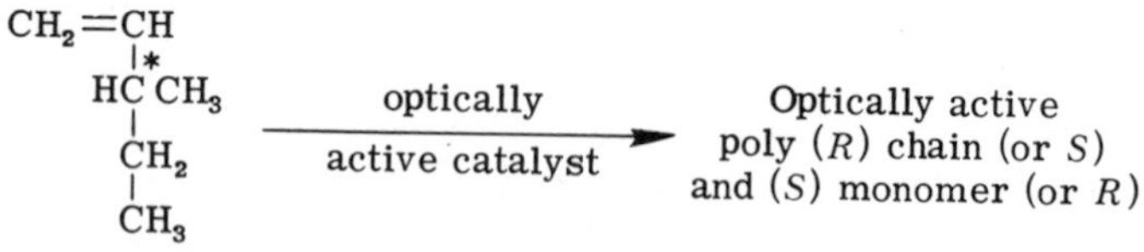

$$CH_2{=}CH$$
$$\overset{|*}{HC}\,CH_3$$
$$\overset{|}{CH_2}$$
$$\overset{|}{CH_3}$$

$\xrightarrow[\text{active catalyst}]{\text{optically}}$

Optically active
poly (R) chain (or S)
and (S) monomer (or R)

(R,S) 3-methyl-1-
pentene

This process is called *stereoelective polymerization*. (The terminology is essentially arbitrary.)

1.6 POLYMERS WITH ASYMMETRIC CENTERS IN THE MAIN CHAIN

It is not possible, as we have seen, for vinyl monomers to have truly asymmetric main chain carbons, but there are other types of monomers, polymerizing in an analogous manner, which do have. A much studied example is propylene oxide (Pruitt and Baggett, 1955; Price and Osgan, 1956; Osgan and Price, 1959; Tsuruta, 1967), which has an asymmetric α-carbon and may be R or S:

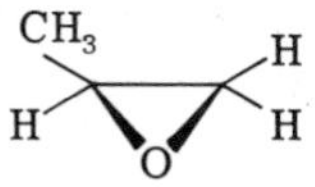

Polymerization may in principle generate four species of configurational triads (shown in planar zigzag projection):

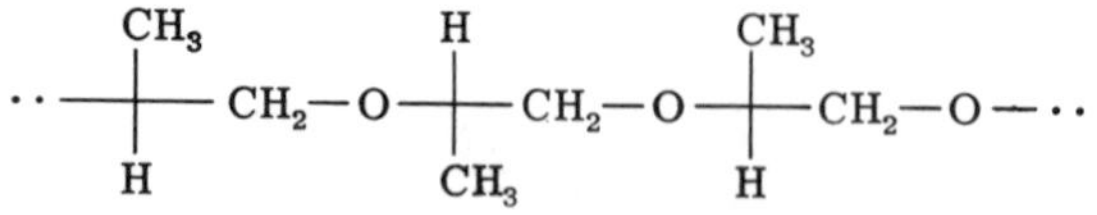

Isotactic, *RRR* or *SSS*

(Isotactic chains may be generated by polymerization of R or S monomer.)

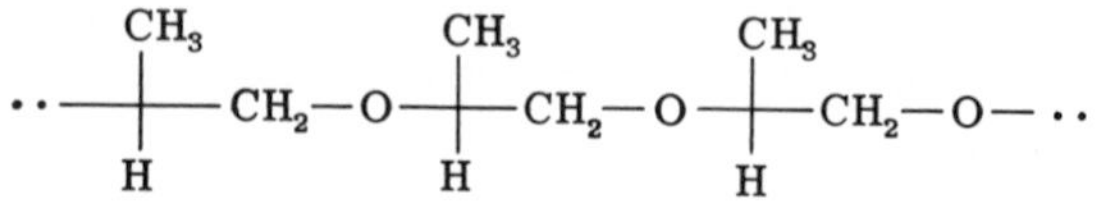

Syndiotactic, *RSR* or *SRS*

In addition to these, two mixed species, termed *heterotactic*, must be recognized:

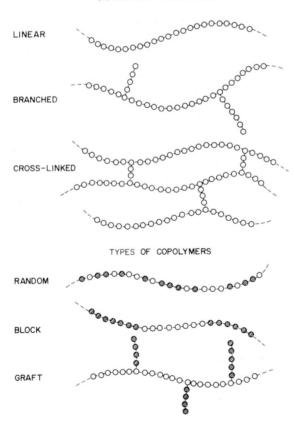

Heterotactic-1 *RRS* or *SSR*

Heterotactic-2 *SRR* or *RSS*

(a) POLYMER STRUCTURES

LINEAR

BRANCHED

CROSS-LINKED

TYPES OF COPOLYMERS

RANDOM

BLOCK

GRAFT

Fig. 1.2. (a) Branched and crosslinked homopolymer structures; (b) copolymer structures.

Because such chains have a *sense of direction*, in contrast to vinyl polymer chains, these latter two units do not superimpose and are not mirror images. Stereoelective and stereoselective polymerization processes may also be applied to such monomers.

There are many other examples of polymer chains formed from chiral monomers, for example,

$$\left[-NH-\overset{\overset{R}{|}}{\underset{*}{CH}}-\overset{\overset{O}{\|}}{C}-\right]_n \qquad \left[-O-\overset{\overset{CH_3}{|}}{\underset{*}{CH}}-\overset{\overset{O}{\|}}{C}-\right]_n$$

Poly-α-amino acids   Polylactic acid

(i)   (ii)

1.7 OTHER TYPES OF ISOMERISM

As shown schematically in Fig. 1.2a, polymer chains are not necessarily linear in form. They may also be *branched* or *cross-linked* into three-dimensional networks. The formation and observation of such structures is discussed in Chapter 6.

Homopolymers consist of only one type of repeating unit, whereas *copolymers* (Fig. 1.2b) are composed of two or more different monomer units. These may be arranged either randomly or with a tendency toward alternation. Copolymers may also possess *block* or *graft* structures, with relatively long sequences of one repeating unit bonded to similar sequences of another. Copolymer structure and its measurement are discussed in Chapter 5.

REFERENCES

Carothers, W. H. (1931). *Chem. Rev.* **8**, 353.
Dall'Asta, G., Mazzanti, G., Natta, G., and Porri, L. (1962). *Makromol. Chem.* **56**, 224.
Flory, P. J. (1953). "Principles of Polymer Chemistry." Cornell Univ. Press, Ithaca, New York.
Harries, C. (1904). *Ber.* **37**, 2708.
Harries, C. (1905). *Ber.* **38**, 1195, 3985.
Huggins, M. L. (1944). *J. Am. Chem. Soc.* **66**, 1991.
Katz, J. R. (1925). *Chem. Z.* **49**, 353.
Mark, H. (1967). "Polymers—Past, Present, Future," *Proc. Robert A. Welch Foundation, Polymers* (W. O. Milligan, ed.), Vol. 10, Chap. 2, Houston, Texas.

Meyer, K. H., and Mark, H. (1928). *Ber.* **61,** 593.

Natta, G. (1955). *J. Am. Chem. Soc.* **77,** 1708.

Natta, G., Farina, M., Donati, M., and Peraldo, M. (1960). *Chim. Ind. (Milan)* **42,** 1363.

Natta, G., Porri, L., and Valenti, S. (1963). *Makromol. Chem.* **65,** 106.

Osgan, M., and Price, C. C. (1959). *J. Polymer Sci.* **34,** 153.

Pino, P., Ciardelli, F., and Lorenzi, G. P. (1963). *J. Polymer Sci., Part C* **4,** 21.

Price, C. C., and Osgan, M. (1956). *J. Am. Chem. Soc.* **78,** 4787.

Pruitt, M. E., and Baggett, J. M. (1955). U.S. Patent 2,706,181.

Schildknecht, C. E., Gross, S. J., Davidson, H. R., Lambert, J. M., and Zoss, A. O. (1948). *Ind. Eng. Chem.* **40,** 2104.

Staudinger, H. (1928). *Ber.* **61B,** 2427.

Staudinger, H. (1932). "Die Hochmolecularen Organischen Verbindungen" Springer-Verlag, Berlin and New York.

Staudinger, H., and Fritschi, J. (1922). *Helv. Chim. Acta* **6,** 705.

Svedberg, T. (1926). *Z. Phys. Chem.* **121,** 65.

Tsuruta, T. (1967). "The Stereochemistry of Macromolecules" (A. D. Ketley, ed.) Marcel Dekker, New York.

Williams, C. G. (1860). *Proc. Roy. Soc. London, Ser. A* **10,** 516.

Chapter 2

THE SPECTROSCOPY
OF MACROMOLECULES

2.1 INTRODUCTION

In this chapter we shall consider the two principal methods by which the structural features of polymer chains discussed in Chapter 1 may be observed and measured: vibrational spectroscopy (infrared and Raman) and nuclear magnetic resonance (NMR) spectroscopy. X-ray diffraction, very powerful for the determination of the structures and conformations of chains regular enough to crystallize, will not be discussed in detail but examples of structures determined by its use will be presented in Chapter 7.

2.2 VIBRATIONAL SPECTROSCOPY

The spectroscopic method that has the longest history for the study of macromolecules is *infrared*. More recently applied and very closely related is *Raman* spectroscopy. (For more complete discussions, see General References at the end of the chapter). Both deal with relatively high frequency processes that involve variation of internuclear distances, i.e., molecular vibration. (Rotational and translational processes will not concern us in polymer spectra.) As a first approximation, let us imagine that these molecular vibrators can be considered as classical *harmonic oscillators*. For a diatomic molecule of unequal masses m_1 and m_2 connected by a bond regarded as a spring with a force constant k, the frequency of vibration expressed in wavenumbers (i.e., cm^{-1} or reciprocal wavelengths) is given by

$$v = (1/2\pi c)(k/m_r)^{1/2} \tag{2.1}$$

where c is the velocity of light and m_r is the *reduced mass*, given by

$$m_r = m_1 m_2/(m_1 + m_2) \quad \text{or} \quad \simeq m_1 \text{ if } m_2 \gg m_1 \tag{2.2}$$

Thus a small mass, such as a hydrogen or deuterium atom, vibrating against a larger one, such as a carbon or chlorine atom, will have essentially the frequency characteristic of the smaller mass. Most molecular vibrational frequencies of interest for polymer characterization will be in the range of 3000 to about 650 cm^{-1}, or in wavelength 2.5 to 15 μm.

Actual molecular vibrators differ from the classical oscillator in two respects. First, the total energy E, expressed in terms of the nuclear displacement x, cannot have any arbitrary value but is confined to discrete energy levels expressed in terms of integral quantum numbers n:

$$E = \left[n + \frac{1}{2}\right]\frac{h}{2\pi}\left[\frac{k}{m_r}\right]^{1/2} = \left[n + \frac{1}{2}\right]hcv \tag{2.3}$$

the transition energies being given by hcv, as shown in Fig. 2.1. Only transitions between adjacent levels are allowed in a quantum mechanical harmonic oscillator. Second, the form of the vibrational potential energy is not parabolic, as in Fig. 2.1, for this would imply infinite bond strength; the true form is *anharmonic*, as shown in Fig. 2.2, the vibrational levels being more closely spaced as n increases. An important consequence of the departure from harmonicity is that the selection rules are relaxed, permitting transitions to levels higher than the next immediately higher one. Transitions from $n = 0$ to $n = 2$ correspond to the appearance of weak but observable *first overtone* bands having slightly less than twice the frequency of the *fundamental* band.

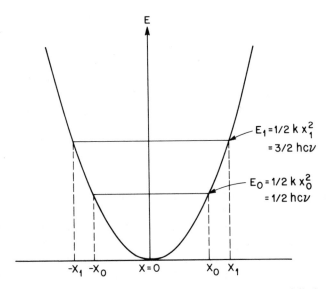

Fig. 2.1. The potential energy of a harmonic oscillator as a function of displacement from equilibrium.

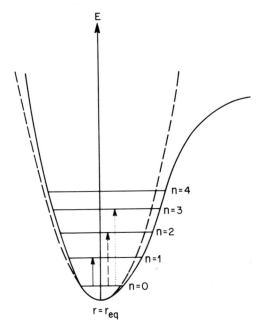

Fig. 2.2. The potential energy and energy levels of an anharmonic oscillator.

2.2.1 Requirements for Infrared and Raman Transitions

The appearance of a vibrational absorption band in the infrared spectral region requires that the impinging radiation supply a quantum of energy ΔE just equal to that of the vibrational transition $hc\nu$. It is also necessary that the atomic vibration be accompanied by a *change in the electric dipole moment of the system*, thus producing an alternating electric field of the same frequency as the radiation field. This condition is often not met, as for example in the vibrations of homopolar bonds such as the carbon–carbon bonds in paraffinic polymers.

The Raman spectrum can give much the same information as the infrared spectrum, but they are in general not identical and can usefully complement each other. In Fig. 2.3 we see at the left the *Rayleigh scattering* process, in which the molecule momentarily absorbs a photon, usually of visible light, and then reradiates it to the ground vibrational state without loss of energy. However, the excited molecule may also return to a higher vibrational state—the next highest in Fig. 2.3 (center)—and the reradiated photon will then be of lower frequency by $\Delta\bar{\nu}$. In a complex molecule there will be many such states, and so the Raman spectrum, like the infrared spectrum,

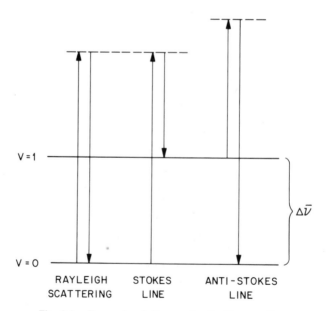

Fig. 2.3. Energy level diagram for the Raman effect.

will appear as a number of lines, called the *Stokes* lines, much weaker than the exciting radiation and appearing shifted to longer wavelength by a few hundred to two or three thousand wavenumbers.

The irradiating light may also excite the molecule from a higher vibrational state ν_1, the emitted photon being then of *higher* frequency than the exciting radiation by $\Delta\bar{\nu}$, and appearing on the short wavelength side of the central line. Such *anti-Stokes* lines will be weaker in intensity than the Stokes lines by the Boltzmann factor expressing the population of state ν_1. The Stokes lines are usually employed for evident practical reasons.

For Raman emission to occur, the *polarizability* of the bond must change during the vibration. For molecules that have a center of symmetry, an exclusion principle applies, stating that transitions that are infrared active will not be Raman active, and vice versa. Polymer molecules generally lack the appropriate center of symmetry and most transitions appear in both types of spectra; an important exception is paraffinic carbon—carbon vibrations, which are inactive in the infrared spectrum, as we have seen, but active in the Raman spectrum.

2.2.2 Types of Vibrational Bands

For polymer molecules (and nonlinear molecules in general) containing N atoms, there will be $3N - 6$ fundamental vibrations. In a complex polymer molecule, the number of transitions might be expected to be too great to deal with, but fortunately this does not happen because great numbers of them are degenerate, allowing us to recognize vibrational bands specific to particular types of bonds and functional groups. These appear in the high-frequency region of the vibrational spectrum at similar positions regardless of the specific compound in which they occur. At the low-frequency end of the spectrum, the vibrational bands are more characteristic of the molecule as a whole; this region is commonly called the "fingerprint" region, since detailed comparison here usually enables specific identification to be made.

In the region near 3000 cm^{-1} appear the *C—H bond stretching vibrations* (Fig. 2.4), which may be asymmetric or symmetric, as illustrated in Fig. 2.5. These occur in nearly all polymer spectra and so are not structurally diagnostic, although useful in a more fundamental sense. At lower frequencies, corresponding to smaller force constants, are the deformation vibrations involving valence angle bending or

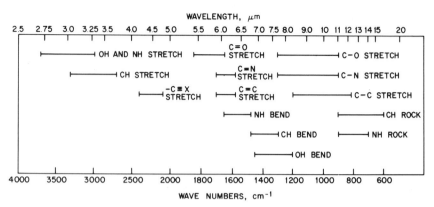

Fig. 2.4.   Infrared bands of interest in polymers.

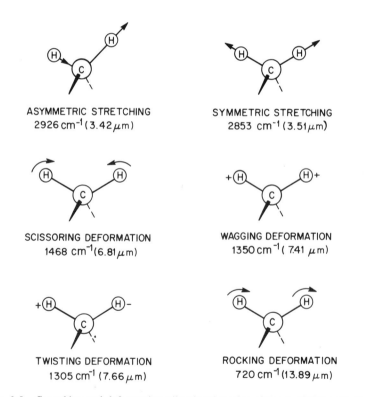

Fig. 2.5.   Stretching and deformation vibrational modes of the methylene group.

scissoring, giving a large band near 1500 cm⁻¹; *wagging* and *twisting* near 1300 cm⁻¹; and finally *rocking* deformations, appearing at the low energy end of the usual spectrum. (At still lower frequencies are torsion and skeletal as well as intermolecular and lattice vibrations, which we shall not discuss here.)

In Fig. 2.4 are also shown a number of other characteristic vibrational bands and their frequency ranges. We may take particular note of the *carbonyl stretch* band near 1700 cm⁻¹, the *C=C stretch* band near 1600 cm⁻¹, and the *olefnic C—H bending* bands between 900 and 1000 cm⁻¹.

2.2.3 Instrumentation and Measurement

In Fig. 2.6 is a schematic diagram of a double beam infrared spectrophotometer. The light source S may be a *globar* (a silicon carbide rod), a *nichrome wire*, or a *Nernst glower*, all electrically heated to at least 1100-1200°C in order for the emitted blackbody radiation to cover the desired frequency range adequately. The *monochromator* G is a rock salt prism or a grating, the wavelength range being swept by rotating the mirror RM. Entrance and exit slits are provided to make the beam reaching the *detector* as monochromatic as possible. The latter may be a Golay cell or a bolometer. The double beam design permits one to blank out absorptions from solvents (if used), from moisture and CO_2 in the air, and in general to provide a flat, stable baseline. The sample and reference beams are compared by moving the comblike attenuator in the reference beam, and recording its movement, until a null is obtained at the detector.

In an infrared spectrometer of the conventional design just described, the spectroscopic information, that is, the intensity of

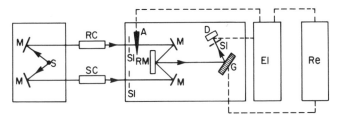

Fig. 2.6. Schematic diagram of a double-beam infrared spectrometer. S = source of radiation; M = mirror; SC = sample cell; RC = reference cell; A = attenuator; RM = rotating mirror; Sl = slit; G = grating or prism; D = detector; El = electronic amplifier; RE = recorder; — = optical path; --- = electrical or mechanical connection (from Kössler, 1967).

radiation as a function of wavenumber, is encoded in the propagation
direction of the dispersed beam. In a spectrometer of *Fourier transform*
design the information is encoded in phase differences and there is no
dispersion of the beam. The optical scheme of the interferometer is
shown in Fig. 2.7.

There are two plane mirrors perpendicular to each other. One (at
the top) is fixed, while the other (at the left) can be moved at a
constant velocity. A beam splitter, consisting of a partially silvered
mirror, is positioned at right angles to both mirrors. The unmodulated
incident beam, typically from a globar (as in a conventional
instrument), enters from the right. A portion of it is reflected to the
fixed mirror and portion passes through the beam splitter to the
moving mirror. Both are reflected back to the beam splitter and pass
through the sample to the detector. Let us assume for the moment
that the entering light is monochromatic. If the two path lengths are
identical, the two beams will be in phase when they return to the beam
splitter and the detector will view a maximum signal. If the movable
mirror is moved by a quarter of a wavelength, the two beams will be
180° out of phase and the detector will see a minimum or zero signal.
If the mirror is moved continuously, the signal will oscillate from
strong to weak for each quarter-wavelength movement of the mirror.

The signal generated by the detector will be a cosine wave or
interferometer beat pattern. When we use all the wavelengths in the
source, with their varied intensities resulting from absorption by the
sample, the output will be the summation of a large number of cosine

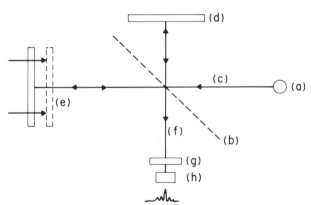

Fig. 2.7. Optical scheme of a Fourier transform infrared spectrometer. (a) source;
(b) beam splitter; (c) unmodulated incident beam; (d) fixed mirror; (e) moving mirror;
(f) modulated exit beam; (g) sample; (h) detector.

functions. These are impossible for the mind to sort out in this form, but can be Fourier transformed from the time domain into the frequency domain and will then appear as a normal spectrum. This will be discussed more fully later in connection with pulsed Fourier transform NMR spectroscopy (Chapter 8). In order to perform the Fourier transform the signal from the detector must be digitized, i.e. converted into binary numbers, and stored in a computer memory. The computer also performs the Fourier transform.

Fourier transform infrared spectroscopy has two chief advantages:

(i)   Sensitivity is much greater. The beam energy is all used without being divided into narrow slices by gratings and slits. A spectrum can be generated in a few seconds. Alternatively, spectra of low intensity can be built up by accumulation in the computer.

(ii) The spectrum, being stored in a computer memory, may be manipulated and operated on at will. Absorption lines may be broadened or narrowed and *difference spectra* may be readily generated.

The requirements of a *Raman spectrometer* are in some respects similar to those of an infrared spectrophotometer, the principal differences being that one uses a monochromatic laser source (typically a helium−neon laser) and that instead of absorption one measures the light scattered at right angles to the sample. A schematic representation is shown in Fig. 2.8.

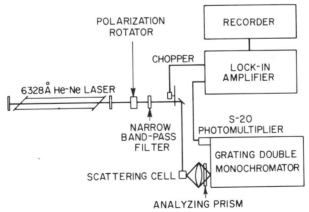

Fig. 2.8. Schematic diagram of a laser-excited Raman spectrometer (from Schaufele, 1970).

Infrared samples are commonly examined in the solid state either as films or as mulls in Nujol or fluorolube. They may also be ground up with KBr, which is transparent to infrared, and observed as pellets. Solutions in CS_2 or CCl_4 are occasionally also used, but for polymers films are most commonly employed. The initial radiation intensity falling on the sample I_0 will be attenuated in proportion to the path length b and, for solutions, to the concentration c; thus

$$I = I_0 e^{-a'bc} \tag{2.4a}$$

where a' is the *extinction coefficient* or *absorbtivity* characteristic of the band observed. For polymer films, c, if expressed in gm cm^{-3}, will be approximately unity. We then have

$$2.303 \log_{10}(I_0/I) = a'bc \tag{2.4b}$$

The quantity $\log_{10}(I_0/I)$ is the *absorbance* or *optical density*, and is the ordinate on the left side of typical infrared spectra.

In Fig. 2.9 are shown the infrared spectra of (a) linear and (b) branched polyethylenes (J. P. Luongo, private communication, 1975). The principal bands are labeled in accordance with the vibrational assignments already discussed. The intense $C—H$ stretch region is twofold owing to the splitting of symmetric and asymmetric bands, as shown in the inset dashed spectrum in (a). Bands for carbonyl groups (1725 cm^{-1}), resulting from slight oxidation, and terminal vinyl groups are observable in the spectrum of the linear polyethylene. If accurate values of a' (from model compounds) and of b are established, the content of these groups can be measured quantitatively. More commonly bands are reported qualitatively as *vw* (very weak), *w* (weak), *m* (medium), *s* (strong), and *vs* (very strong).

The measurement of the branch content of polyethylene from the infrared spectrum in Fig. 2.9 is discussed in Section 6.3.1.

The Raman spectra of polymers resemble the infrared spectra. In crystalline paraffinic hydrocarbons, bands appear below 800 cm^{-1} corresponding to accordionlike vibrations of the chains, which are in a planar zigzag conformation. In polyethylene chains, similar bands may be observed for motions of chain folds.

Vibrational bands that are Raman active but infrared inactive are those corresponding to C—C stretch; in crystalline $n—C_{44}H_{90}$ (and in

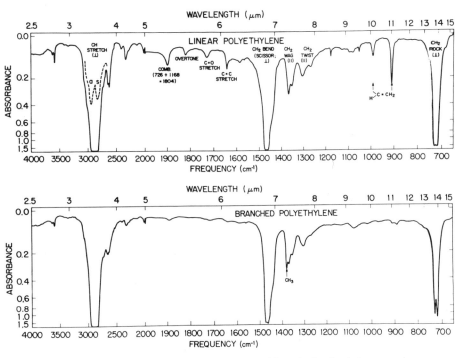

Fig. 2.9. Infrared spectra of linear and branched polyethylenes.

polyethylene) these appear at 1140 and 1060 cm^{-1}, as shown in Fig. 2.10, which shows the Raman Stokes lines (H. N. Cheng and A. P. Ginsberg, private communication, 1975). In general, while infrared is more useful for identifying polar substituents on a polymer chain, Raman spectroscopy is more powerful in characterizing the homonuclear polymer backbone. It also can be used for very small samples, particularly those that cannot conveniently be made into films or mulls, since it is only necessary to fill the focused laser beam, which may be only 10 μm in diameter.

An important technique, long used for the study of solid polymer samples, is the measurement of *infrared dichroism*. The radiation is polarized by a small rotatable grid of fine parallel wires placed in the beam. If the transition moment of the infrared band, i.e., the movement of the electrons accompanying the vibration, is parallel (or has a component parallel) to the electric vector of the advancing wave, absorption will be strong; if perpendicular, absorption will be weak. In Fig. 2.11 is shown in a schematic fashion the absorption of a polarized wave train when the planar zigzag chain of polyethylene is oriented in

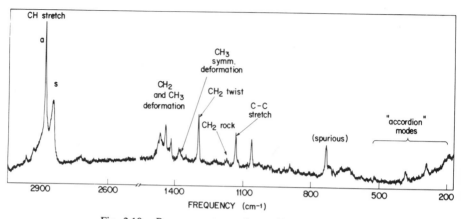

Fig. 2.10.   Raman spectrum of crystalline n-$C_{44}H_{90}$.

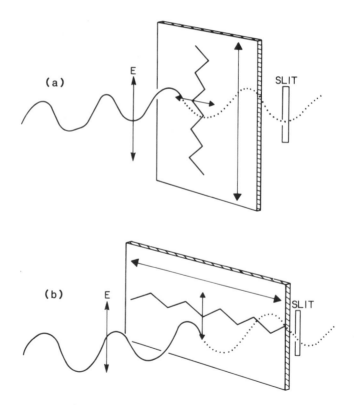

Fig. 2.11.   Dichroism in infrared rocking absorption in polyethylene.

such a manner that the methylene rocking vibration of the planar zigzag chain is (a) perpendicular and (b) parallel to the polarization of the beam (J. P. Luongo, private communication, 1975). In crystalline polyethylene the rocking band at 725 cm^{-1} is actually split into two bands corresponding to the two possible phase relationships of its motion to that of the neighboring chains. When the beam is unpolarized, the rocking band for an extruded sample of polyethylene is as shown in Fig. 2.12a. The *dichroic ratio* is defined as $I_\perp/I_{\parallel}$, the ratio of the absorption when the direction of orientation of the sample is perpendicular to the beam polarization to that when it is parallel. When polyethylene is drawn or extruded the chains tend to become oriented in the draw direction. The dichroism of the rocking band does not show the idealized behavior suggested by Fig. 2.11, but the 730 cm^{-1} band does show a dichroic ratio *less* than one, and is termed a parallel or π band, whereas the 720 cm^{-1} band shows a dichroic ratio markedly *greater* than one, as expected from Fig. 2.11, and is termed a perpendicular or σ band. Dichroism is observed in nearly all the bands of oriented polymer samples and gives very useful information concerning chain conformation and morphology.

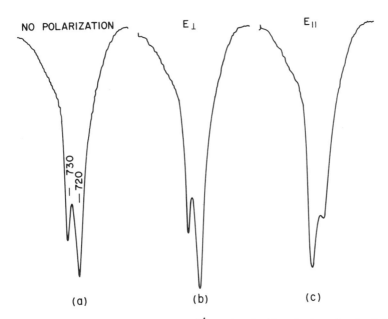

Fig. 2.12   Dichroism in the 720/730 cm^{-1} rocking doublet of oriented polyethylene (J. P. Luongo, private communication, 1975).

2.3 NUCLEAR MAGNETIC RESONANCE SPECTROSCOPY

2.3.1 The Basic Phenomenon

The phenomenon of nuclear magnetic resonance depends on the fact that some nuclei possess *spin* or angular momentum. (See General References under Nuclear Magnetic Resonance Spectroscopy at the end of the chapter.) Such nuclei are described by spin quantum numbers I (usually referred to simply as "the spin") having integral or half-integral values. When placed in a magnetic field of strength B_0, (commonly expressed in teslas, the S.I. unit of magnetic flux density; 1 tesla is equivalent to 10,000 gauss) such nuclei occupy quantized magnetic energy levels the number of which is equal to $2I+1$ and the relative populations of which are normally given by a Boltzmann distribution. We shall deal only with nuclei for which $I=1/2$: the proton (^{1}H) and the ^{13}C nucleus. These nuclei have two magnetic energy states separated by

$$\Delta E = h \nu_0 = 2\mu B_0 \qquad (2.5)$$

in which μ is the magnetic moment of the nucleus. Transitions between these energy levels can be made to occur by means of a resonant radio frequency (rf) field B_1 having a frequency ν_0. In Fig. 2.13, Eq. (2.5) is plotted for the proton; the resonant rf frequency is shown for six different magnetic fields employed in current spectrometers. (The fields above *ca.* 2.5 tesla require superconducting solenoid magnets.) For the ^{13}C nucleus, which has a magnetic moment one-fourth that of proton, the resonant frequencies will be one-fourth of those indicated. The resonance condition for any nucleus may be alternatively expressed as

$$\omega_0 = \gamma B_0, \qquad (2.6)$$

where ω_0 is the frequency expressed in radians-s^{-1}, equal to $2\pi\nu_0$, and γ is the *magnetogyric ratio*, equal to $2\pi\mu/Ih$. The use of magnetogyric ratios has the advantage that resonant frequencies are always directly proportional to γ, regardless of the spin.

In Fig. 2.14 is shown a basic schematic diagram of a conventional continuous wave (cw) spectrometer. The sample is contained in the tube A to which the rf field is applied by means of coil B wrapped about the sample. A magnetic field is applied in a direction

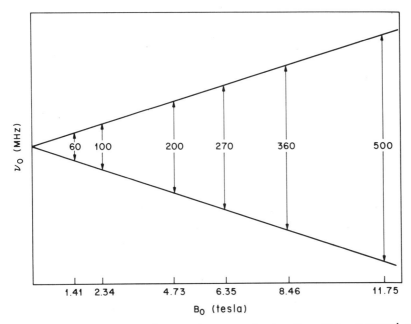

Fig. 2.13. The splitting of magnetic energy levels of protons, expressed as resonance frequency ν_0 in varied magnetic field B_0, expressed in teslas.

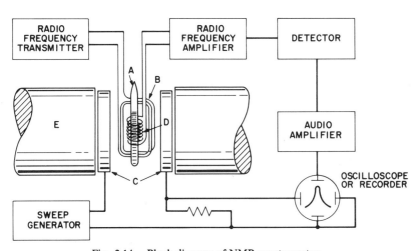

Fig. 2.14. Block diagram of NMR spectrometer.

perpendicular to the axis of the coil and, by means of the sweep coils C, is slowly increased ("swept") until resonance occurs. At this point energy is absorbed from the rf field and this absorption may be detected and recorded. Alternatively, the nuclear moments as they are turned over at resonance induce a voltage in a second coil D, and this voltage is amplified and recorded. (It is also equally feasible to hold B_0 constant and develop the spectrum by sweeping ν_0.)

2.3.2 The Chemical Shift

The value of NMR to the chemist lies in the fact that at any particular value of ν_0, all nuclei of a given species—say, all protons—do not resonate at exactly the same value of B_0 (or vice versa). Equation (2.5) requires a modification, which is small in terms of relative magnitudes but very significant, because resonance actually occurs at slightly different values of B_0 for each type of proton, depending upon its chemical binding and position in the molecule. The cause of this variation is the cloud of electrons about each nucleus, which tends to shield the nucleus against the magnetic field, thus requiring a slightly higher B_0 to achieve resonance than for a bare proton. Protons attached to or near electronegative groups such as OR, OH, OCOR, CO_2R, and halogens experience a lower density of shielding electrons and resonate at lower B_0 (or at a higher value of ν_0 if B_0 is held constant). Protons removed from such groups, as in hydrocarbon chains, resonate at higher B_0. Similar structural relationships are observed for ^{13}C nuclei. These variations are termed *chemical shifts* and are commonly expressed in relation to tetramethylsilane (TMS) as the zero of reference. The total range of variation of proton chemical shifts in organic compounds is only of the order of 10 ppm. For ^{13}C nuclei (in common with all other magnetic nuclei), it is much greater—over 200 ppm; this is a principal reason for the intense interest in the study of polymers by ^{13}C NMR despite certain inherent observing difficulties (see below). For any nucleus, the separation of chemically shifted resonances, expressed in hertz, is proportional to B_0. An important advantage of high-field magnets is the greater resolution of peaks and finer discrimination of structural features of polymer chains that they make possible.

2.3.3 Nuclear Coupling

Another important parameter in NMR spectra is *nuclear coupling*. Magnetic nuclei may transmit information to each other concerning their spin states through the intervening covalent bonds. If a nucleus has n sufficiently close, equivalently coupled neighbors, its resonance will be split into $n + 1$ peaks, corresponding to the $n + 1$ spin states of the neighboring group of spins. Intensities are given by simple statistical considerations and are therefore proportional to the coefficients of the binomial expansion. Thus, one neighboring spin splits the observed resonance to a doublet, two produce a 1:2:1 triplet, three a 1:3:3:1 quartet, and so on. The strength of the coupling is denoted by J and is expressed in hertz (Hz). The coupling of protons on adjacent saturated carbon atoms, termed *vicinal* coupling, varies with the dihedral angle ϕ, (a) trans couplings where $\phi = 180°$, being substantially larger than (b) gauche couplings, where $\phi = 60°$.

H

H

H

H

(a)   (b)

J_{gauche} is typically $2-4$ Hz and J_{trans} ranges from 8 to 13 Hz. These considerations are of great importance in studying the conformations of polymer chains (Chapter 7). The dependence of J on ϕ is generally well described by the so-called Karplus relationship, shown in Fig. 2.15, although the magnitude of the coupling depends somewhat on the nature of the substituents on the bonded carbon atoms.

The chemical shifts and J couplings of nuclei in polymers do not essentially differ from those of small molecules of analogous structure. The spectrum of ethyl orthoformate, $CH(OCH_2CH_3)_3$ (Fig. 2.16a), shows a single peak for the formyl proton and a quartet and triplet for the CH_2 and CH_3 protons, respectively, of the ethyl group, in accordance with the rules just described. The spectrum of poly(vinyl ethyl ether) (Fig. 2.16b) shows corresponding resonances and, in addition, those of the β protons appearing at 1.5 ppm and of the α protons under the CH_2 quartet. Chemical shifts are expressed in ppm from tetramethylsilane, which appears at 0.0. The broadening of the lines in (b) is in part due to the slower motions of the large molecule in solution. [In the solid state, where motion is very slow, NMR lines are commonly so broad as to obscure all structural information (see

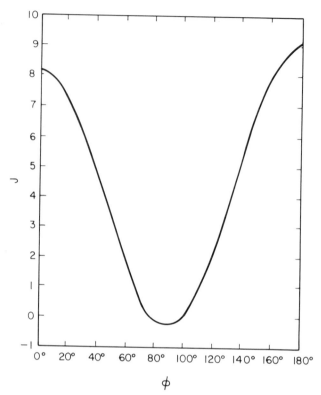

Fig. 2.15. *J* coupling of two protons, expressed in hertz, as a function of dihedral angle ϕ.

Chapter 8)]. The broadening is also in part due to structural complexities to be dealt with in Chapter 3.

2.3.4 Carbon 13 Spectroscopy

The observation of *carbon spectra* presents special problems because the magnetic isotope ^{13}C has an abundance of only 1.1% and also, as we have seen, a relatively small magnetic moment—about one-fourth that of the proton. The resulting decrease in observing sensitivity can be compensated by use of the *pulsed Fourier transform* technique, combined with spectrum accumulation, in a manner analogous to that described for infrared spectroscopy in Section 2.2. In this method, the rf field is supplied as a periodic pulse of a few microseconds duration and of much greater power than used in conventional cw spectroscopy. The pulse sets all the carbon nuclei over the usual chemical shift range

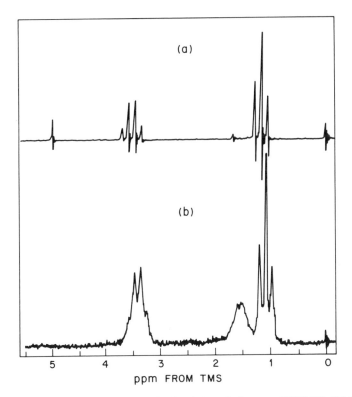

Fig. 2.16.   The NMR spectrum of (a) ethyl orthoformate $HC(OCH_2CH_3)_3$, and (b) poly(vinyl ethyl ether), both observed at 60 MHz in 15% solutions in carbon tetrachloride.

into resonance at the same instant and obviates the need for sweeping the field or frequency, thus saving a factor of at least 100 in observing time.   The time-domain spectra, appearing as interferograms following each pulse, are added in a computer; after enough are accumulated for an acceptable signal-to-noise ratio, the summed spectra are transformed to the frequency domain, i.e., to the usual form.   This method will be more fully described in Chapter 8.

Nuclear coupling between ^{13}C nuclei and directly bonded protons is strong (125-250 Hz).   The resulting multiplicity is often helpful in carbon assignments but is usually abolished by *double resonance*, i.e., by providing a second rf field tuned to the proton resonance frequency. The resulting multiplet collapse and accompanying *nuclear Overhauser enhancement* [see Levy and Nelson, 1972; Stothers, 1972; Wehrli and Wirthlin, 1976; and Chapter 8] give a striking improvement in signal-to-noise ratio.

GENERAL REFERENCES

I   Vibrational Spectroscopy
General

Colthup, N. B., Daly, L. H., and Wiberly, S. E. (1964). "Introduction to Infrared and Raman Spectroscopy." Academic Press, New York.
Potts, W. J., Jr. (1963). "Chemical Infrared Spectroscopy," Vol. I, Technique. Wiley, New York.

Polymers

Hendra, P. J. (1969). Laser-Raman spectra of polymers, *Adv. Polymer Sci.* **6**, 151.
Hummel, D. O. (1966). "Infrared Spectroscopy of Polymers." Wiley (Interscience), New York (Elementary).
Kössler, I. (1967). Infrared-absorption spectroscopy, *In* "Encyclopedia of Polymer Science and Technology" (N. M. Bikales, ed.). Wiley (Interscience), New York (Elementary).
Krimm, S. (1960). Infrared spectra of high polymers, *Adv. Polymer Sci* **2**, 51 (moderately advanced).
Krimm, S. (1968). Infrared spectra and polymer structure, *Pure Appl. Chem.* **16**, 369 (advanced).
Schaufele, R. F. (1970). *Macromol. Rev.* **4**, 67.
Zbinden, R. (1964). "Infrared Spectroscopy of High Polymers." Academic Press, New York (advanced).

II   Nuclear Magnetic Resonance
General

Becker, E. D. (1980). "High Resolution NMR," 2nd ed. Academic Press, New York.
Bovey, F. A. (1969). "Nuclear Magnetic Resonance Spectroscopy." Academic Press, New York.
Emsley, J. W., Feeney, J., and Sutcliffe, L. H. (1966). "High Resolution Nuclear Magnetic Resonance" Pergamon, Oxford.
Levy, G. C., and Nelson, G. L. (1972). "Carbon-13 Nuclear Magnetic Resonance for Organic Chemists." Wiley (Interscience), New York.
Stothers, J. B. (1972). "Carbon-13 NMR Spectroscopy." Academic Press, New York.
Wehrli, F. W., and Wirthlin, T. (1976). "Interpretation of Carbon-13 NMR Spectra" Heyden, New York.

Polymers

Bovey, F. A. (1972). "High Resolution NMR of Macromolecules." Academic Press, New York.
Randall, J. C. (1977). "Polymer Sequence Determination. Carbon-13 NMR Method" Academic Press, New York.
Schaefer, J. (1974). *In* "Topics in Carbon-13 NMR Spectroscopy" (G. C. Levy, ed.), Chapter 4. Wiley (Interscience), New York.

Chapter 3

STEREOCHEMICAL CONFIGURATION AND ITS OBSERVATION

3.1 INTRODUCTION

Of the various forms of isomerism that we have seen to be possible in vinyl and diene polymer chains, the relative configurations of pseudoasymmetric centers are of special significance. It also happens

39

that this form of isomerism is particularly amenable to study by NMR. No other form of spectroscopy approaches the discrimination of NMR in this respect. Only X-ray diffraction is of comparable power and we shall later discuss some of the results obtained by its use. But X-ray diffraction of course requires that the polymer be at least partially crystalline and most polymers are not.

It is now well recognized that the physical and mechanical properties of vinyl (and some diene) polymers are critically dependent upon their stereochemical configuration. For example, isotactic poly(methyl methacrylate) has a crystalline melting point of 160°, whereas the syndiotactic polymer melts at 200°. The glass temperature, i.e., the temperature at which a rather abrupt transition between the glassy and rubbery states takes place, occurs at about 45° for the isotactic polymer and at 115° for the syndiotactic polymer, the chains of which are evidently much stiffer.

Before considering in detail the spectroscopy of vinyl polymers, let us first introduce the subject through the discussion of *model compounds*.

3.2 MODEL COMPOUNDS FOR VINYL POLYMER CHAINS

The power of NMR to determine the configurations of vinyl polymer chains arises from its sensitivity to equivalence or nonequivalence of similarly bonded nuclei. To amplify this statement we must look more closely at what we mean by "equivalence."

The lowest order of equivalence is simply fortuitous equivalence of chemical shift within the experimental resolution possible under the conditions employed. In structural studies of polymers, where linewidths tend to be broad, such equivalence degrades the amount of information available and is often misleading. Fortuitous degeneracy of chemical shifts can often be removed by a change of solvent or by an increase in magnetic field strength.

The next level of equivalence is equivalence by reason of *molecular symmetry*. If a molecule has an element of symmetry, such as a translation axis, a twofold (or higher) rotation axis, or a mirror plane, then those magnetic nuclei that exchange their positions upon performing the appropriate symmetry operation have symmetry equivalence and must have the same chemical shift. In polymer chains formed of identical repeating units and long enough so that end effects may be ignored, hundreds or thousands of nuclei may be equivalent in this sense. On the other hand, in non-repeating structures such as protein molecules, few of the nuclei have symmetry equivalence, although substantial groups of them may have chemical shift equivalence.

The highest level of equivalence is termed *magnetic equivalence*. A group of spins A_n is said to have magnetic equivalence if all n spins have the same chemical shift and are equally coupled to each of the m spins in every other group of magnetically equivalent spins in the molecule to which appreciable coupling occurs. Thus, the protons in ethane form a single equivalent group A_6, while those in ethanol form two equivalent groups (ignoring the hydroxyl proton):

$$CH_3 - CH_2 + OH)$$

$$B_3 \quad\ A_2$$

Related situations can occur in polymer proton or fluorine spectra; CH_3 and CF_3 groups represent what are commonly considered magnetically equivalent groups of spins, although they usually do not rigorously conform to the definition. In general, we shall not need to observe this distinction and may term nuclei "equivalent" even if they have only symmetry equivalence. For ^{13}C spectra, such considerations are unimportant.

Let us now consider some less obvious situations. First, consider a molecule of the general formula:

$$
\begin{array}{c}
M \\
| \\
X - C - Y \\
| \\
M
\end{array}
$$

The M groups may be observable nuclei such as H, F, or $^{13}CH_3$, or they may be equivalent groups of spins such as CH_3 or CF_3. If the groups X and Y have a plane or axis of symmetry—for example, phenyl, methyl, or halogen—the groups M will be equivalent and will have the same chemical shift. In ethylbenzene, the methylene protons are equivalent:

This seems straightforward. However, if one of the groups has no element of symmetry the M groups will be nonequivalent and will exhibit (at least in principle) different chemical shifts. Although we shall not be primarily concerned with amino acids and polypeptides in this discussion, the amino acid valine provides a convenient example to illustrate this point, the α−carbon being asymmetric:

$$
\begin{array}{ccc}
CH_3 & & CO_2^- \\
| & & | \\
H-\underset{\beta}{C} & \!\!\!-\!\!\!-\!\!\!- & \underset{\alpha}{C}-H \\
| & & | \\
CH_3 & & NH_3^+
\end{array}
$$

In Fig. 3.1 is shown the 100 MHz proton spectrum of L-valine in neutral D_2O solution. Note that two methyl doublets appear near

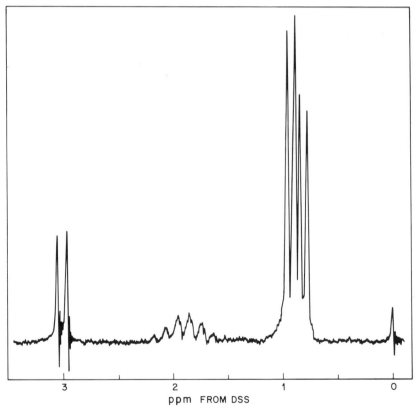

Fig. 3.1. The 100 MHz ^1H spectrum of valine (zwitterion form in D_2O, $pD \simeq 6$; DSS is sodium 2,2-dimethyl-2-silapentane-5-sulfonate.)

1.0 ppm. Why are the methyl groups differentiated? Consider the staggered conformers of L-valine about the C_α—C_β bond:

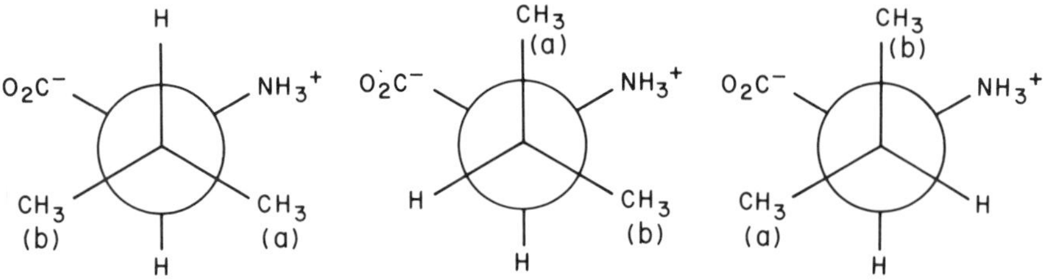

In each of these conformers, and in any possible conformer, the environments of methyl (a) and methyl (b) are always nonequivalent. Rapid rotation about the C_α—C_β bond does not cause averaging of their environments because methyl (a) never experiences the environment of methyl (b) and vice versa. The appearance of two methyl doublets is not the result of slow rotation. "Freezing out" the conformers by lowering the temperature to the point where equilibration is slow on the NMR time scale would indeed cause a change in the spectrum, since we would then see a separate spectrum for each conformer; there would then be six separate methyl proton doublets or ^{13}C singlets. (In general, the rotational barriers in polymer chains are not high enough to enable one to freeze out the conformers in the dissolved state, although this can occur in the solid state; see Chapter 8.)

A similar example is provided by the methylene protons of an ethyl group attached to an asymmetric center; under these circumstances, a differentiation of these protons is observed (i, below):

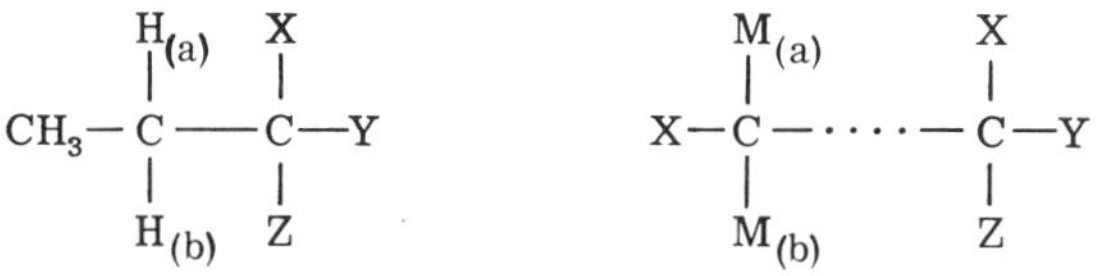

It is not necessary for the chiral center to be directly bonded to the carbon bearing the M groups, although the effect will generally be largest when this is the case. In principle, it could be removed by several intervening bonds (ii). Furthermore, it is not required that the molecule be chiral as a whole. Thus, the group Y may have the structure

$$Y = \begin{array}{c} P \quad M \\ | \quad | \\ -\!\!\!-\!\!\!-C-X \\ | \quad | \\ Q \quad M \end{array}$$

so that the overall structure is

$$X-\overset{\displaystyle M_{(a)}}{\underset{\displaystyle M_{(b)}}{C}}-\!\!\!-\!\!\!-\overset{\displaystyle P}{\underset{\displaystyle Q}{|}}-\!\!\!-\!\!\!-\overset{\displaystyle M_{(a)}}{\underset{\displaystyle M_{(b)}}{C}}-X$$

The molecule as a whole is not asymmetric nor are there any asymmetric centers. Nevertheless, it fulfills the conditions for differentiation of $M_{(a)}$ and $M_{(b)}$. Many examples of this sort are now known and the subject has been reviewed (van Gorkom and Hall, 1968).

Moving a step closer to vinyl polymers, consider molecules of the structure*

$$CH_3-\overset{\displaystyle R}{\underset{\displaystyle H}{C^*}}-\overset{\displaystyle H}{\underset{\displaystyle H}{C^*}}-\overset{\displaystyle R}{\underset{\displaystyle H}{C^*}}-CH_3$$

Here there are two similar chiral centers. If they are mirror images, d and l or R and S, the molecule is meso and is not asymmetric or optically active, having a plane of symmetry. In either Fischer or planar zigzag projection (see the Appendix to this chapter) it is:

$$CH_3-\overset{\displaystyle R}{\underset{\displaystyle H}{\underset{d}{|}}}-\overset{\displaystyle H_{(a)}}{\underset{\displaystyle H_{(b)}}{|}}-\overset{\displaystyle R}{\underset{\displaystyle H}{\underset{l}{|}}}-CH_3$$

---

* Note that molecular structures written in full as here do not imply any particular handedness for the chiral centers. Relative handedness is implied, however, in the projection formulas.

The protons $H_{(a)}$ and $H_{(b)}$ will be nonequivalent from previous arguments. However, if the asymmetric centers are of like chirality d,d (R,R) or l,l (S,S) we have a pair of enantiomers that give identical NMR spectra (except in a chiral solvent):

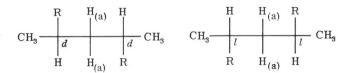

These molecules, although asymmetric, have a twofold axis of symmetry and are sometimes termed "dissymmetric" to distinguish them from chiral molecules having no element of symmetry. Because of the twofold axis, the methylene protons $H_{(a)}$ and $H_{(b)}$ are equivalent and exhibit a single chemical shift.

The 100 MHz NMR spectra of the methine and methylene protons of meso and racemic *(dl)*-2,4-diphenylpentane, i.e., with R=phenyl in the above structures, are shown in Fig. 3.2. It is to be observed that whereas the methylene protons appear as a triplet at a single chemical shift in the racemic model spectrum, the more complex meso methylene spectrum corresponds to two proton chemical shifts separated by 0.21 ppm. The differentiation of methylene protons is observed in isotactic polymers as well, as we shall see, and serves as an *absolute measure of tacticity*, no appeal to X-ray results being necessary. In syndiotactic polymers, for which the racemic enantiomers serve as models, the methylene protons are equivalent and will give a single chemical shift. Extensive studies have been made by both NMR and IR of model 2,4-disubstituted pentanes in which R is phenyl, chloro, bromo, cyano, acetoxy (CH_3COO-), trifluoroacetoxy, hydroxyl, carboxyl, carbomethoxyl, and CH_3O.

It is important to try to avoid confusion about the designation of the asymmetric (or pseudoasymmetric) centers. They have been appropriately designated as d and l (or R and S) in this discussion of model compounds, but problems can arise in transferring this terminology to polymer chains. Different authors have adopted different conventions. It is common (but not universal) to regard isotactic chains as composed of monomer units of like chirality, $\cdots dddd \cdots$ or $\cdots llll \cdots$, yet the "isotactic" model compound is meso, i.e. *dl*. A syndiotactic polymer chain is thought of as having asymmetric centers of alternating chirality, yet the model compounds are *dd* or *ll*.

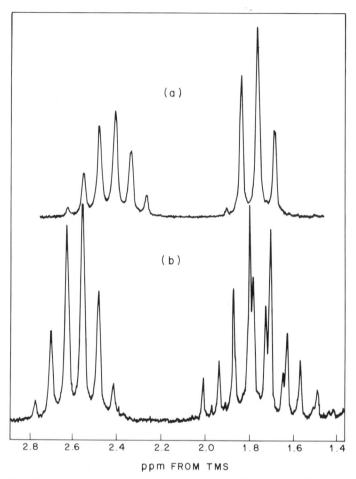

Fig. 3.2.   The 100 MHz spectra of (a) racemic- and (b) meso- 2,4-diphenylpentane,
10% (v/v) in chlorobenzene at 35°.

The 2,4-disubstituted pentanes are models of *dyads* of monomer
units, which we shall henceforth designate as *m* (for meso) and *r* (for
racemic).   The 2,4,6-trisubstituted heptanes are models of *triads* of
monomer units and are designated by the terms we shall employ in
describing polymer chains.   (Note that the isotactic model is again
meso but that the syndiotactic model is also meso; the "heterotactic"
model compound is *dl* or racemic.)

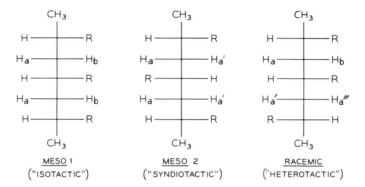

MESO 1
("ISOTACTIC")

MESO 2
("SYNDIOTACTIC")

RACEMIC
("HETEROTACTIC")

3.3 POLYMER CHAINS

3.3.1 Poly(methyl methacrylate)

It is now well recognized that these symmetry considerations can be extended to polymer chains with hundreds or thousands of pseudoasymmetric centers. In one way, very long chains simplify matters. Ignoring the chain ends, we may now consider $H_{(a)}$ and $H_{(a')}$, not strictly equivalent in the syndiotactic model heptane above, as entirely equivalent in a long purely syndiotactic chain.

It is instructive to begin with an α,α'-disubstituted polymer, poly(methyl methacrylate). The spectrum is in this case simplified by the absence of vicinal coupling of main-chain protons:

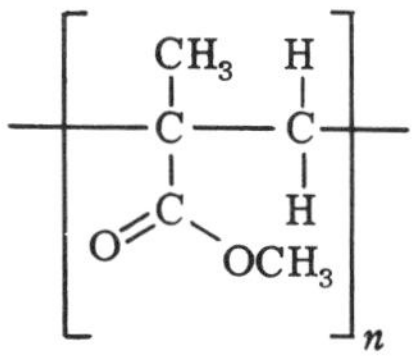

Geminal coupling of the methylene protons is present, however, and when they are nonequivalent it is manifested as an AB quartet. The 60 MHz proton spectra of two methyl methacrylate polymers are shown in Fig. 3.3. Spectrum (a) is that of a polymer prepared with a free radical initiator; spectrum (b) represents a polymer prepared with an anionic initiator, phenylmagnesium bromide, in a hydrocarbon solvent (Bovey and Tiers, 1960). We may surmise from the marked

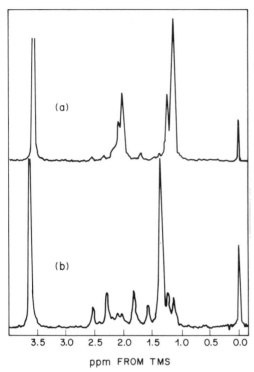

Fig. 3.3. The 60 MHz spectra of 15% (w/v) solutions in chlorobenzene of poly(methyl methacrylate) prepared with (a) a free radical initiator and (b) with an anionic initiator, phenylmagnesium bromide (Bovey and Tiers, 1960).

differences between these spectra that the nature of the initiator has a profound influence on the stereochemistry of the polymer chain. These are more than "fingerprint" differences and may be interpreted in detail. Let us first consider the chain in terms of monomer dyads, of which we have seen there are two types, meso *(m)* and racemic *(r)*:

meso, m   *racemic, r*

The racemic dyad, as we have seen for the model pentane, has a twofold symmetry axis, if we disregard for the moment the effect of neighboring units, and therefore the methylene protons are equivalent and will appear as a singlet resonance. The meso dyad has no symmetry axis and so $H_{(a)}$ and $H_{(b)}$ are nonequivalent. In Fig. 3.3, the methylene resonance, centered near 2.2 ppm, is principally a singlet although complicated by other resonances. It thus appears that the free radical polymer is predominantly syndiotactic, but that isotactic sequences occur with observable probability. The methylene spectrum of the anionically initiated polymer is mainly an AB quartet, but there are other resonances indicative of residual syndiotactic sequences. The spacing of the AB quartet corresponds to a geminal coupling of ~-15 Hz, which is exceptionally large for such a structure and probably indicates an abnormally small $\overset{H}{\underset{}{}}\diagdown_{C}\diagup^{H}$ angle accompanying a spreading of the main-chain $\overset{C}{\diagdown}_{C_\beta}\diagup^{C}$ angle owing to steric crowding. We thus find, as intimated earlier, that proton NMR can provide an absolute measure of the stereochemical configuration of a vinyl polymer chain without recourse to X-ray diffraction. These observations have since been extended to a large number of other polymers, of which we shall discuss a few further examples.

More detailed, but not absolute, stereochemical information can be obtained from the α—methyl resonances. (The ester methyl resonance is not sensitive to configuration.) The smallest meaningful sequence is the monomer triad, corresponding to the 2,4,6-trisubstituted heptane model compounds we have discussed in Section 3.2:

Isotactic, *mm*

Syndiotactic, *rr*

Heterotactic, *mr* (or *rm*)

These sequences are appropriately designated in terms of *m* and *r*, as shown. Measurement of the relative intensities of the *mm*, *mr* and *rr* α—methyl resonances, which, from what has already been said, must appear from left to right in this order in both spectra (a) and (b), gives a valid statistical representation of the structure of each polymer.

From the α−methyl triad resonances, we may gain considerable insight into the mechanism of propagation. Let us designate by P_m the probability that the polymer chain will add a monomer unit to give the same configuration as that of the last unit at its growing end, i.e., that an m dyad will be generated. We assume that P_m is independent of the configuration of the growing chain. In these terms the generation of the chain is a Bernoulli trial process. It is like reaching into a large jar of balls marked m and r and withdrawing a ball at random. The proportion of m balls in the jar is P_m. The probability of an r dyad is $1 - P_m$. Since two monomer additions are required to form a triad sequence, it can be readily seen that the probabilities of forming mm, mr, and rr triads are given by

$$[mm] = P_m^2 \qquad\qquad (3.1)$$

$$[mr] = 2P_m(1-P_m) \qquad\qquad (3.2)$$

$$[rr] = (1-P_m)^2 \qquad\qquad (3.3)$$

The heterotactic sequences are given double statistical weighting because both directions mr and rm must be counted. A plot of these relationships is shown in Fig. 3.4. It will be noted that the proportion of mr (heterotactic) units rises to a maximum at $P_m = 0.5$, corresponding to a strictly random or atactic configuration, for which the proportion $[mm]:[mr]:[rr]$ will be 1:2:1. For any given polymer, if Bernoullian, the mm, mr, and rr sequence frequencies, as estimated from the relative areas of the α−methyl resonances, would lie on a single vertical line in Fig. 3.4 corresponding to a single value of P_m. Spectrum (a) in Fig. 3.3 corresponds to these simple statistics, P_m being 0.24 ± 0.01. The polymer corresponding to (b) does not. The propagation statistics are in this case more complex and can be interpreted to indicate that the probability of isotactic propagation is not independent of the stereochemical configuration of the propagating chain (see later discussion). Free radical and cationic propagations always give predominantly syndiotactic chains; anionic initiators may also do so if strongly complexing ether solvents such as dioxane or glycol dimethyl ether are employed rather than hydrocarbon solvents as in Fig. 3.3b.

Vibrational spectra also reveal stereochemical differences. In Fig. 3.5 are shown infrared spectra of films of the same predominantly syndiotactic (top) and isotactic (bottom) methyl methacrylate polymers

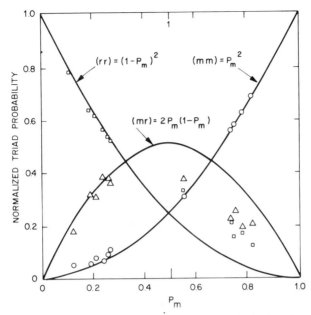

Fig. 3.4. The probabilities of isotactic *(mm)*, heterotactic *(mr)*, and syndiotactic *(rr)* triads as a function of P_m, the probability of *m* placement. The points on the left-hand side are for methyl methacrylate polymers prepared with free radical initiators and those on the right-hand side for polymers prepared with anionic initiators.

as in Fig. 3.3. It is evident that, in addition to other smaller differences, there is a conspicuous band at 1060 cm^{-1} in the syndiotactic polymer spectrum that is entirely absent in that of the isotactic polymer. This band can serve as a quick measure of the chain stereochemistry, but in general infrared is not as discriminating nor as quantitative as NMR.

3.3.2 Propagation Statistics

In Fig. 3.6 is shown the building up of a vinyl polymer chain by (a) a Bernoulli trial process and by (b) a first-order Markov process. In Bernoulli trial propagation, the chain end is not represented as having any particular stereochemistry, neither *m* nor *r*. The outcome of the choice does not, in other words, depend on the outcome of any previous choice. It is sometimes stated that the addition is influenced only by the end unit of the growing chain. This is true in that the mode of addition (*m* or *r*) may be biased by the steric requirements of the chain end and the approaching monomer. Indeed, Bernoulli trial processes may be and often are biased; free radical polymers are

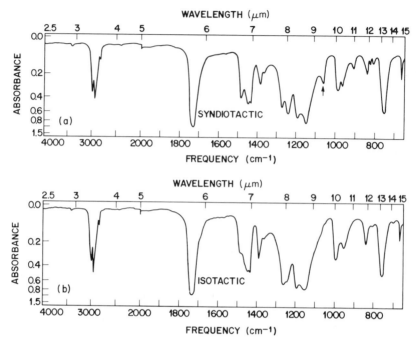

Fig. 3.5. Infrared spectra of predominantly (a) syndiotactic and (b) isotactic films of polymethyl methacrylate.

usually predominantly syndiotactic although Bernoullian. But such statements should not be understood to imply that the mode of addition is influenced by the *stereochemistry* of the growing chain*.

The first-order Markov sequence is generated by propagation steps in which the mode of addition of the approaching monomer *is* influenced by whether the growing chain-end is *m* or *r*, as shown in Fig. 3.6. The choice thus depends on the outcome of the previous choice. We now have four probabilities characterizing the stereochemistry of the propagation process, $P_{m/m}$, $P_{m/r}$, $P_{r/m}$, $P_{r/r}$, defined as in Fig. 3.6. These probabilities are not independent of each other, however, for we have the conservative relationships:

$$P_{m/m} + P_{m/r} = 1 \qquad (3.4)$$

---

* It is probable that in free radical propagation, and possibly anionic and cationic as well, the growing chain end is actually trigonal, i.e. "flat", and that its stereochemical configuration is decided as the next monomer adds. Such considerations do not affect the formal statistical arguments.

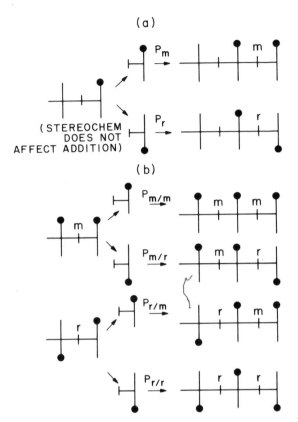

Fig. 3.6. Schematic representation of (a) Bernoulli trial and (b) first-order Markov propagation processes.

$$P_{r/r} + P_{r/m} = 1 \qquad (3.5)$$

There are therefore actually only two independent probabilities, which we choose as $P_{m/r}$ and $P_{r/m}$. It can be shown that:

$$P_{m/r} = \frac{(mr)}{2(mm)+(mr)} \quad [=(1-P_m) \text{ if Bernoullian}] \qquad (3.6)$$

$$P_{r/m} = \frac{(mr)}{2(rr)+(mr)} \quad [=P_m \text{ if Bernoullian}] \qquad (3.7)$$

Thus, if the configuration conforms to a Bernoullian model, $P_{m/r}$ and $P_{r/m}$, calculated according to Eqs. 3.6 and 3.7, will total unity. For a free radical poly(methyl methacrylate), typical findings are:

$$(mm) = 0.04; \quad (mr) = 0.36; \quad (rr) = 0.60$$

$$P_{m/r} = \frac{0.36}{0.08+0.36} = 0.82$$

$$P_{r/m} = \frac{0.36}{1.20+0.36} = 0.23.$$

The sum of $P_{m/r}$ plus $P_{r/m} = 1.05$, and therefore this polymer is Bernoullian within experimental error. Applying this test to an anionically initiated polymer [polymer (b) in Fig. 3.2], we have:

$$P_{m/r} = \frac{0.14}{1.50+0.14} = 0.085$$

$$P_{r/m} = \frac{0.14}{0.22+0.14} = 0.39$$

Here, $P_{m/r}$ plus $P_{r/m} = 0.47$. The quantity $(P_{m/r}+P_{r/m})$ may vary from zero to 2. If both $P_{m/r}$ and $P_{r/m}$ are zero, the polymer is composed of infinitely long blocks of isotactic and syndiotactic structures. A mechanical mixture of isotactic and syndiotactic chains would be indistinguishable spectroscopically from such a structure. A tendency toward a blocklike structure is indicated when $(P_{m/r}+P_{r/m}) < 1$. If $P_{m/r}$ and $P_{r/m}$ are both unity, the chain is heterotactic: $\cdots mrmrmrmr \cdots$. No such structure has been observed. A tendency in this direction is indicated when $(P_{m/r}+P_{r/m}) > 1$.

A simpler statistical approach is implicit in the preceding discussion. Consider the following "block" chain:

Defining the configurational blocks as shown, the isotactic sequence of n_m monomer units contains n_m-1 *mm* triads and n_m *m* dyads, while the syndiotactic sequence contains n_r-1 *rr* triads and n_r *r* dyads. The heterotactic resonance is an inverse measure of "blockiness":

$$(mr)^{-1} = \frac{1}{2} (\bar{n}_m + \bar{n}_r) = \bar{n} \qquad (3.8)$$

where $\bar{n}_m$ and $\bar{n}_r$ are number average lengths of isotactic and syndiotactic blocks, respectively, and $\bar{n}$ is the number average length of all blocks. It can also be shown that

$$\bar{n}_m = 1 + \frac{2(mm)}{(mr)} \qquad (3.9)$$

$$\bar{n}_r = 1 + \frac{2(rr)}{(mr)} \qquad (3.10)$$

3.3.3 Longer Configurational Sequences

Through advances in the design of NMR spectrometers, particularly increased magnetic field strength, one may readily observe configurational sequences longer than triads. In Table 3.1 are shown planar zigzag projections of such sequences, together with their frequency of occurrence as a function of the probability of meso placement, P_m, assuming Bernoullian propagation. Again it should be noted that all unsymmetrical sequences receive double statistical weighting because they must be counted in both directions, directionality not being experimentally distinguishable. The tetrads—and all "even-ads"—refer to observations of β—methylene carbons and protons (or other substituents), while the "odd-ads" refer to α—carbons or substituents. Figure 3.7 shows a plot of tetrad sequence probabilities as a function of P_m, i.e. of the appropriate relationship in Table 3.1; Fig. 3.8 shows the corresponding relationships for pentad sequences. Since *m* and *r* are conjugate terms in these relationships, the frequency plots are symmetrical, as implied in the legends to these figures.

Resonances for tetrad sequences should appear as fine structure in the dyad spectra, while pentad sequences should appear as fine structure on the triad resonances. In Fig. 3.9 are shown the

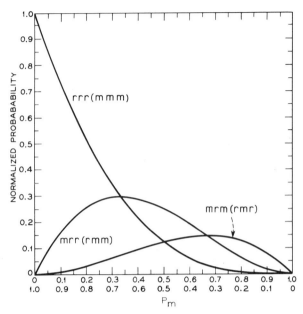

Fig. 3.7. Tetrad sequence probabilities as a function of P_m. For *rrr*, *mrr(rrm)*, and *mrm*, the upper P_m scale is used; for *mmm*, *rmm(mmr)*, and *rmr*, the lower P_m scale is used.

β—methylene proton spectra of predominantly syndiotactic (a) and predominantly isotactic (b) poly(methyl methacrylate), the same polymers as shown in Fig. 3.2 but now observed at 220 MHz (Frisch *et al.*, 1968). The splitting of the "singlet" resonance in (a), alluded to previously, is now very clear. The same peaks can be seen, in altered relative intensity, in (b). In the latter, the predominant resonances are those of the *mmm* quartet. Smaller quartet resonances are observable in both spectra. It may be noted that *r*-centered tetrads, e.g. *mrr*, do not necessarily appear as singlets. In Fig. 3.10 are shown the α—methyl proton spectra of the same samples. The predominant *rr* triad resonance in (a) (at *ca.* 1.1 ppm) is now clearly split into pentad fine structure, which also appears as shoulders on the *mr* and *mm* resonances. Corresponding resonances appear in spectrum (b), but with an observable downfield shift. [This downfield shift may arise from the difference in predominant chain conformation between polymers (a) and (b) (Flory and Baldeschwieler, 1966); we shall discuss these matters in more detail in Section 7.8.]

TABLE 3.1

	α Substituent			β-CH$_2$		
	Designation	Projection	Bernoullian probability	Designation	Projection	Bernoullian probability
Triad	Isotactic, mm (i)		P_m^2	Dyad: meso, m	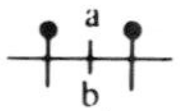	P_m
	Heterotactic, mr (h)	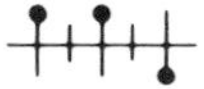	$2P_m(1-P_m)$	racemic, r	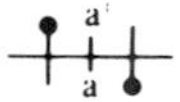	$(1-P_m)$
	Syndiotactic, rr (s)		$(1-P_m)^2$ 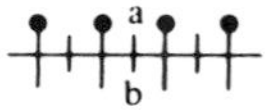			
Pentad	$mmmm$ (isotactic)		P_m^4 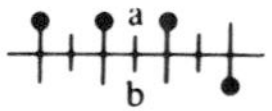	Tetrad: mmm		P_m^3
	$mmmr$		$2P_m^3(1-P_m)$ 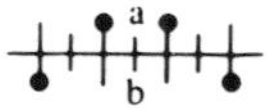	mmr		$2P_m^2(1-P_m)$
	$rmmr$		$P_m^2(1-P_m)^2$ 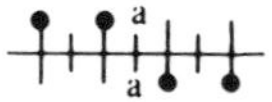	rmr		$P_m(1-P_m)^2$
	$mmrm$		$2P_m^3(1-P_m)$	mrm		$P_m^2(1-P_m)$
	$mmrr$		$2P_m^2(1-P_m)^2$	rrm		$2P_m(1-P_m)^2$
	$rmrm$ (heterotactic)		$2P_m^2(1-P_m)^2$	rrr	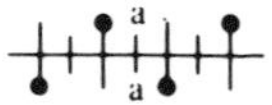	$(1-P_m)^3$
	$rmrr$		$2P_m(1-P_m)^3$			
	$mrrm$	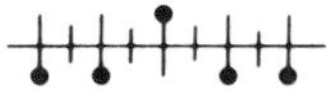	$P_m^2(1-P_m)^2$			
	$rrrm$	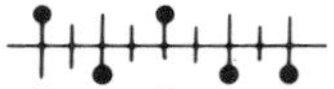	$2P_m(1-P_m)^3$			
	$rrrr$ (syndiotactic)		$(1-P_m)^4$			

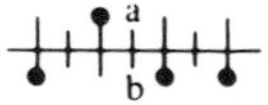

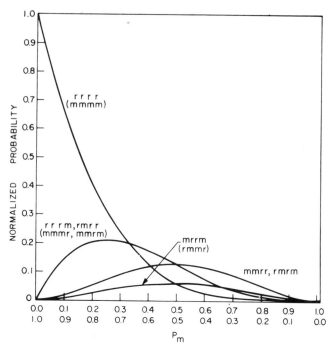

Fig. 3.8. Pentad sequence probabilities as a function of P_m. The lower scale should be used for sequences in parentheses.

3.3.4 Numbers and Necessary Relationships of Configurational n-ads

The number of observationally distinguishable configurational sequences or $n-$(ads), i.e. the number of types of sequences containing n monomer units, is designated $N(n)$ and obeys the following relationship (Frisch *et al.*, 1966):

n	2	3	4	5	6	7	8	9
$N(n)$	2	3	6	10	20	36	72	136

or, in general

$$N(n) = 2^{n-2} + 2^{m-1},$$

where $m = n/2$ if n is even and $m = (n-1)/2$ if n is odd. It is evident that discrimination at the heptad level will present problems, although

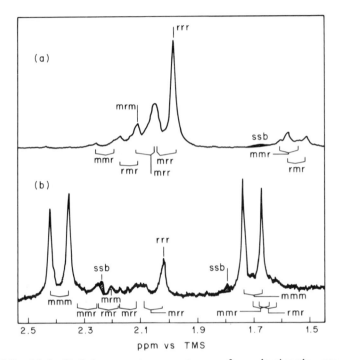

Fig. 3.9. (a) β—Methylene proton spectrum of predominantly syndiotactic poly(methyl methacrylate) observed at 220 MHz (10% w/v solution in chlorobenzene at 135°.) (b) β—Methylene proton spectrum of predominantly isotactic poly(methyl methacrylate), same conditions. The vertical markers for quartet resonances are joined by slanting lines which meet at the chemical shift. "Ssb" denotes spinning sidebands.

we shall later consider cases where this has proved possible. Discrimination of longer sequences will not usually prove to be possible, and is also unlikely to be important.

The probabilities of occurrence of configurational sequences are governed by certain necessary relationships which are based on elementary considerations and are entirely independent of the statistics generated by any particular mechanism (Frisch et al., 1966). Some of these relationships are presented in Table 3.2. Being independent of mechanism, they can often serve as useful tests of peak assignments. Any set of assignments requiring peak intensity ratios in conflict with the appropriate relationships in Table 3.2 cannot be correct. (It should be understood that the intensities are normalized over each species of n—ads.)

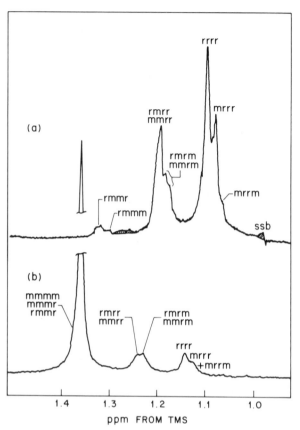

Fig. 3.10. (a) $\alpha-$Methyl proton spectrum of predominantly syndiotactic poly(methyl methacrylate) and (b) predominantly isotactic poly(methyl methacrylate). Polymers and conditions are the same as in Fig. 3.9.

3.4 THE EFFECT OF POLYMERIZATION TEMPERATURE ON STEREOCHEMICAL CONFIGURATION

The dependence of stereochemical configuration on the temperature of polymerization has a significant bearing on our understanding of the mechanism of propagation and is readily amenable to measurement by NMR. If we express the propagation steps in terms of absolute reaction rate theory, the rate constant for isotactic or m placement of monomer units is

$$k_m = (kT/h)\exp\left[\left[\Delta S_m^{\ddagger}/R\right] - \left[\Delta H_m^{\ddagger}/RT\right]\right] \qquad (3.11)$$

TABLE 3.2

Some Necessary Relationships Among Sequence Frequencies

Dyad	$(m) + (r) = 1$
Triad	$(mm) + (mr) + (rr) = 1$
Dyad–Triad	$(m) = (mm) + \frac{1}{2}(mr)$
	$(r) = (rr) + \frac{1}{2}(mr)$
Triad–Tetrad	$(mm) = (mmm) + \frac{1}{2}(mmr)$
	$(mr) = (mmr) + 2(rmr) = (mrr) + 2(mrm)$
	$(rr) = (rrr) + \frac{1}{2}(mrr)$
Tetrad–Tetrad	sum $= 1$
	$(mmr) + 2(rmr) = 2(mrm) + (mrr)$
Pentad–Pentad	sum $= 1$
	$(mmmr) + 2(rmmr) = (mmrm) + (mmrr)$
	$(mrrr) + 2(mrrm) = (rrmr) + (rrmm)$
Tetrad–Pentad	$(mmm) = (mmmm) + \frac{1}{2}(mmmr)$
	$(mmr) = (mmmr) + 2(rmmr) = (mmrm) + (mmrr)$
	$(mrr) = \frac{1}{2}(mrmr) + \frac{1}{2}(rmrr)$
	$(mrm) = \frac{1}{2}(mrmr) + \frac{1}{2}(mmrm)$
	$(rrm) = 2(mrrm) + (mrrr) = (mmrr) + (rmrr)$
	$(rrr) = (rrrr) + \frac{1}{2}(mrrr)$

The rate constant for syndiotactic or r placement is

$$k_r = (kT/h)\exp\left[\left[\Delta S_r^{\ddagger}/R\right] - \left[\Delta H_r^{\ddagger}/RT\right]\right] \qquad (3.12)$$

From these relations we have

$$P_m = \frac{k_m}{k_m + k_r} \qquad (3.13a)$$

$$= \frac{\exp\left[-\left[\Delta G_m^{\ddagger} - \Delta G_r^{\ddagger}\right]/RT\right]}{1 + \exp\left[-\left[\Delta G_m^{\ddagger} - \Delta G_r^{\ddagger}\right]/RT\right]} \qquad (3.13b)$$

and

$$k_m/k_r = P_m/(1-P_m)$$

$$= \exp\left\{\left[\left(\Delta S_m^{\ddagger}-\Delta S_r^{\ddagger}\right)/R\right]-\left[\left(\Delta H_m^{\ddagger}-\Delta H_r^{\ddagger}\right)/RT\right]\right\} \quad (3.14)$$

The differences in activation enthalpies and entropies for isotactic and syndiotactic propagation are given by

$$\Delta H_m^{\ddagger} - \Delta H_r^{\ddagger} = \Delta\left(\Delta H_p^{\ddagger}\right) = -\left.Rd\ln\left[P_m/(1-P_m)\right)\right]d(1/T) \quad (3.15)$$

$$\Delta S_m^{\ddagger} - \Delta S_r^{\ddagger} = \Delta\left(\Delta S_p^{\ddagger}\right) = R\ln\left[P_m/(1-P_m)\right] + \Delta\left(\Delta H_p^{\ddagger}\right)/T \quad (3.16)$$

This treatment of course implies that the configuration statistics are Bernoullian and can therefore be described by a single probability P_m. If this were not the case, interpretation would be considerably more complicated and has not been attempted. All studies so far reported have been confined to free radical systems. We present data for methyl methacrylate here, and in Section 3.6.2 we consider the behavior of vinyl chloride and vinyl bromide.

Measurements of P_m for free radical methyl methacrylate propagation have been obtained over a 178° polymerization temperature range by Bovey (1960), Fox and Schnecko (1963), and Otsu et al. (1966). Figure 3.11 shows an Arrhenius plot of these data according to Eq. (3.15). From this plot it is found that for methyl methacrylate

$$\Delta H_m^{\ddagger} - \Delta H_r^{\ddagger} = \Delta\left(\Delta H_p^{\ddagger}\right) = 1 \text{ kcal } -\text{mol}^{-1}$$

$$= 4.2 \text{ kJ } -\text{mol}^{-1}$$

$$\Delta S_m^{\ddagger} - \Delta S_r^{\ddagger} = \Delta\left(\Delta S_p^{\ddagger}\right) = 1 \text{ cal } -\text{mol}^{-1}-\text{deg}^{-1}$$

$$= 4.2 \text{ J } -\text{mol}^{-1}-\text{deg}^{-1}$$

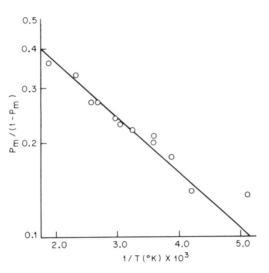

Fig. 3.11. Arrhenius plot of the stereochemical configuration of free radical poly(methyl methacrylate).

In the system, syndiotactic placement is thus favored by the additional enthalpy of activation required for isotactic placement, which itself is favored by entropy. It is commonly observed, as here, that P_m decreases with decreasing polymerization temperature.

3.5 MORE COMPLEX PROTON SPECTRA; COMPUTER SIMULATION

The proton spectra of vinyl polymers having only one α−substituent are considerably more complex than that of poly(methyl methacrylate) owing to vicinal coupling of main-chain protons, and are often quite unintelligible at first glance. Multiplet spacings and intensities do not necessarily obey the simple "binomial" relationships described in Section 2.3.3, a general observation when the J couplings and chemical shift differences of the coupled protons are comparable in magnitude. Successful interpretation usually requires computer simulation. To do this, one must adopt a spin system that adequately represents a polymer chain. Since only a limited number of spins can be accommodated, a long chain presents something of a problem. A model pentane or heptane, such as we have discussed earlier (Section 3.2), is not a satisfactory solution as it does not adequately represent the vicinal coupling of the α−protons to their β−proton neighbors. A satisfactory spin model is provided by imagining that the polymer chain is turned back on itself to form a cyclic "dimer" (Chûjô *et al.*, 1962; Tincher, 1962). This of course should not be taken too literally as it is only intended to be a spin bookkeeping system. For an isotactic polymer we have the six-spin system:

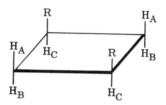

For a syndiotactic chain:

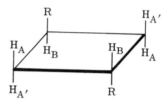

Only vicinal and geminal couplings are included, four-bond couplings being ignored, as they are very small.

We now consider some examples of the calculation of such spectra.

3.5.1 Poly(isopropyl acrylate)

In Fig. 3.12a is shown the 100 MHz main-chain proton spectrum of a predominantly isotactic poly(isopropyl acrylate) prepared with a Grignard initiator (Schuerch *et al.*, 1964; Heatley and Bovey, 1968a). Figure 3.12b shows a computed spectrum that closely matches the observed spectrum, but requires the presence of *ca.* 5% of syndiotactic sequences, the α–protons of which appear as a multiplet centered at 2.58 ppm, while the β–protons are centered at 1.84 ppm. In Fig. 3.12b a linewidth of 2.4 Hz is assumed. Figure 3.12c shows the "stick" spectrum, i.e., with zero linewidth. Figure 3.12d shows the spectrum of "atactic" polymer prepared with a free radical initiator.

The spectrum of the free radical polymer is markedly different from that of the isotactic polymer, although the α–proton chemical shift is not very sensitive to tacticity. The principal β–proton intensity is centered near 1.82 ppm, close to that of the *r* dyad in the Grignard initiated polymer, indicating that the polymer is predominantly syndiotactic, as expected. The spectrum gives the

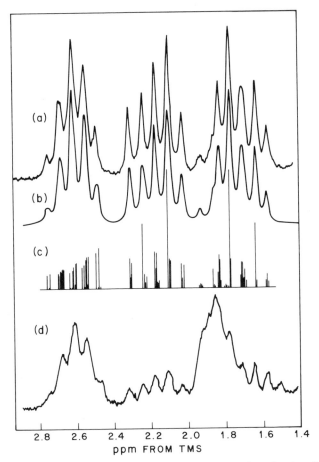

Fig. 3.12. The 100 MHz main-chain proton spectrum of predominantly isotactic poly(isopropyl acrylate): (a) observed; (b) calculated with parameters given in the text; (c) "stick" spectrum corresponding to (b); and (d) "atactic" polymer. Experimental spectra were obtained on 10% solutions in chlorobenzene at 140° (Heatley and Bovey, 1968a).

impression of poorer resolution but this is actually the result of overlapping tetrad resonances.

The computer simulation of the isotactic polymer does not address the question of the nature of the syndiotactic "defects" but assumes the same parameters for an r dyad regardless of whether it occurs in a mrm, mrr, or possibly even rrr tetrad. This is not quite correct, as (among other things) it may be that H_A and $H_{A'}$ are not equal in chemical shift in the *mrr* sequence. In the treatment of poly(vinyl chloride) (Section 3.5.2) we shall elaborate the calculation to deal with this point.

The parameters employed in the computer simulation are as follows:

δ_A :   1.69 ppm
δ_B :   2.14 ppm
δ_C :   2.58 ppm
J_{AB} = −14.0 Hz
J_{AC} =   6.0 Hz
J_{BC} =   7.5 Hz

δ_A :   1.84 ppm
δ_B :   1.58 ppm
$J_{AA'}$ = −14.0 Hz
$\frac{1}{2}(J_{AB} + J_{AB'})$ = 7.0 Hz

3.5.2 Poly(vinyl chloride)

The efforts to understand fully the structure of this polymer have extended over a long period of time. This is to be attributed in part to its complex chemistry, which we shall deal with extensively in later chapters, but mainly to its great economic importance. It is very widely used in films, molding, and electrical insulation, although it has the drawback of inherent chemical instability (see Section 6.3.2).

In early work on the structure of poly(vinyl chloride), no consideration was given to the question of stereochemical configuration, as this concept did not enter into the consciousness of those few organic chemists who concerned themselves with polymers. In the period from *ca.* 1950 onward, as infrared spectroscopy was increasingly applied to the study of polymer structure, there was established a fairly settled conclusion that poly(vinyl chloride), as normally prepared by free radical initiation, is syndiotactic (Krimm *et al.*, 1958, 1959; Krimm, 1960). The X-ray diffraction study of Natta and Corradini (1956) did not actually support this conclusion as strongly as the infrared spectroscopists appeared to assume. Natta and Corradini, in commenting on the imperfect layer lines of the fiber diagram of a specimen of free radical poly(vinyl chloride), stated that the "lack of order along the *c*-axis can be attributed to imperfections in the syndiotactic arrangement of the Cl atoms." In fact it was clear, even from very early NMR studies (Johnsen, 1961; Bovey and Tiers, 1962), that poly(vinyl chloride) is nearly atactic, with P_m of *ca.* 0.42 for material prepared at temperatures near 50°C.

Figure 3.13a shows the 60 MHz spectrum of poly(vinyl chloride) (Bovey *et al.,* 1963). It appears to consist of a pentuplet at *ca.* 4.5 ppm for the $\alpha-$protons and another (non-binomial) pentuplet at *ca.* 2.1 ppm corresponding to the $\beta-$protons. The nature of these multiplets is more clearly revealed by use of two important techniques. The first of these is *decoupling* (sometimes termed "double resonance"), mentioned in Section 2.3.3. To take the simplest case, suppose that we have an *AX* spin system (two doublets), and that while observing *A* with the usual rf field B_1 we simultaneously provide a second, much stronger rf field B_2 tuned to the resonant frequency of *X*. We find that the *A* doublet collapses to a singlet. The field B_2 has caused the magnetic moment of nucleus *X* to jump between its spin states so rapidly that it is no longer observably coupled to *A*. Reversing the experiment, we can cause the collapse of the *X* doublet by irradiating *A*. When both nuclei are of the same species this is termed *homonuclear decoupling.* The elimination of coupling

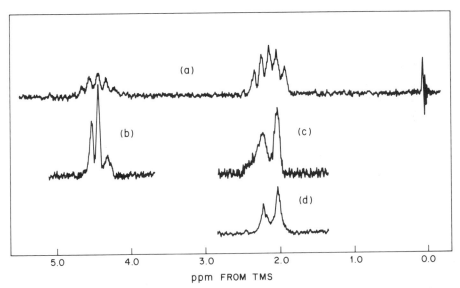

Fig. 3.13. The 60 MHz proton spectra of poly(vinyl chloride), observed in 15% (w/v) solution in chlorobenzene. (a) Normal spectrum (at 170°); (b) $\alpha-$protons upon irradiation of $\beta-$protons (150°), (c) $\beta-$protons upon irradiation of $\alpha-$protons (150°); (d) poly(α-d_1-vinyl chloride).

multiplicity, leaving only chemical shift information, can be very helpful. Thus spectrum (b) in Fig. 3.13 shows the appearance of the α-proton multiplet upon strong irradiation at the center of the β-proton multiplet; three singlets are revealed. In spectrum (c), the β-protons are collapsed to two broad "singlets" by irradiation of the α-protons. Assuming the upfield, somewhat larger resonance to correspond to r methylene dyads, one may cause selective collapse of the α-resonances by careful adjustment of the decoupling frequency and in this way show that the three decoupled α-proton peaks correspond to rr, mr, and mm triads in order of increasing shielding.

Spectrum (d) illustrates another very useful method of spectral simplification. The replacement of a proton by deuterium removes its resonance from the spectrum, since the magnetogyric ratio (Section 2.3.1) of deuterium (or ^2H) is only about one-sixth of that of ^1H. More important, the corresponding scalar couplings are reduced in the same ratio, and since vicinal couplings in polymer chains are about 6 Hz (see Section 3.5.1) the multiplet spacings are reduced to the same magnitude as the linewidth. Figure 3.13d shows the spectrum of poly(α-d_1-vinyl chloride)

$$\left[CH_2 - \overset{\overset{\displaystyle Cl}{\displaystyle |}}{CD} \right]_n$$

prepared from the α-deutero monomer. As expected it resembles spectrum (c) but close examination reveals that it is not merely two singlets (the m resonance should be an AB quartet) and that a dyad analysis is inadequate. This becomes very obvious when this polymer is observed at 220 MHz (Heatley and Bovey, 1969). In Fig. 3.14, the spectrum now reveals much detail; at least nine resonances are clearly resolved. Below the experimental spectrum are shown a set of tetrad AB quartets and singlets that give a good match; their intensities correspond to a P_m of 0.43. The geminal coupling is taken as -15 Hz. The mrm and rrr tetrads are singlets, as might be expected; but it is curious that mmm is nearly so, while the r-centered tetrad mrr shows widely different chemical shifts.

It would be more satisfactory if the nondeuterated polymer could be analyzed, and this in fact can be done. In Fig. 3.15a is shown the 220 MHz spectrum of the β-protons of poly(vinyl chloride), essentially the same material as that employed in Fig. 3.13. The chemical shifts and geminal couplings are the same as used for

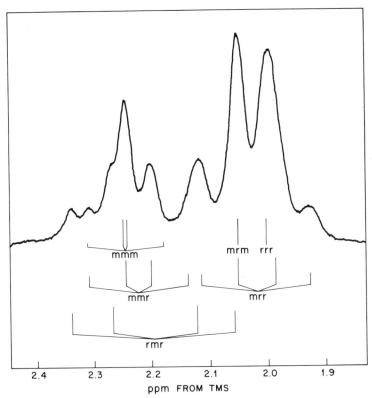

Fig. 3.14. The 220 MHz proton spectrum of poly(α-d_1-vinyl chloride), observed in 15% (w/v) solution in chlorobenzene at 140°.

Fig. 3.14. For the *m*-centered tetrads, the vicinal couplings giving the best match were $J_{AC} = 6.5$ Hz and $J_{BC} = 7.5$ Hz, similar to those found for poly(isopropyl acrylate) (see Section 3.5.1). The rather large linewidth of 5.0 Hz gives considerable margin for error in these values. For the *r*-centered tetrads, corresponding vicinal couplings were 2.0 and 11.0 Hz and a linewidth of 7.0 Hz was employed. The tetrad subspectra were added together with intensities corresponding to a Bernoullian prescription (again with $P_m = 0.43$) to give (b), which closely matches the observed spectrum.

The 220 MHz α—proton spectrum was also simulated by combining ten pentad subspectra with the same P_m and vicinal couplings and with chemical shifts consistent with the decoupled spectrum described previously. The result is shown in Fig. 3.16. The limits of error are fairly wide here.

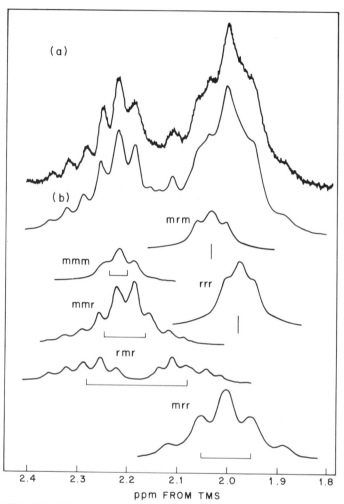

Fig. 3.15.   The (a) 220 MHz $\beta-$proton spectrum of poly(vinyl chloride), observed in 5% solution in chlorobenzene at 140°; (b) calculated spectrum, composed of the six component tetrad subspectra shown below. Horizontal bars show chemical shifts in each tetrad.

The analysis of the proton spectra of stereoirregular polymers in this manner is clearly a considerable task and has therefore only rarely been carried out.   Carbon-13 spectroscopy (Section 3.6) is more straightforward and is now the preferred method.

3.5.3 Polystyrene

The earliest reported polymer spectra were of polystyrene (Saunders and Wishnia, 1958; Odajima, 1959; Bovey *et al.,* 1959). In

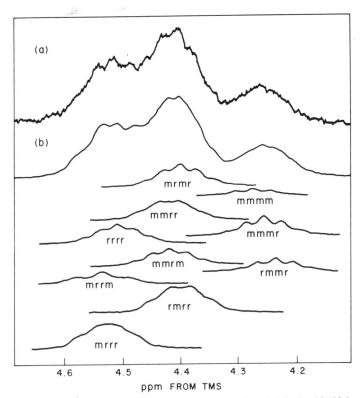

Fig. 3.16. The 220 MHz α-proton spectrum of poly(vinyl chloride), same conditions as for Fig. 3.15.

Fig. 3.17 (top) is shown the 60 MHz proton spectrum of isotactic polystyrene, prepared with a Ziegler—Natta catalyst (Bovey *et al.*, 1965). Spectrum (a) at low field represents the aromatic protons; it is notable that the *ortho* protons appear as a distinct and separate resonance upfield from the *meta-para* resonance. This arises from the enforced close proximity of the phenyl groups, the *ortho* protons being in the ring-current shielding region of the neighboring aromatic rings. (Waugh and Fessenden, 1957, 1958; Johnson and Bovey, 1958). Spectrum (d) is that of the α- and β-protons, in order of increasing shielding. At this observing frequency, the β-methylene protons have the same chemical shift within experimental error (± 0.05 ppm).

This ambiguity is resolved at 220 MHz. In Fig. 3.18 spectra (a) and (d) are those of the main-chain protons of the same polymer observed in *o*-dichlorobenzene at 130° (Heatley and Bovey, 1968b). On the left are the observed and calculated spectra of the α-protons;

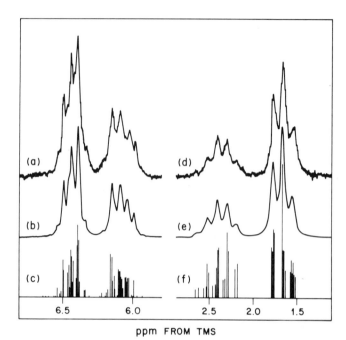

Fig. 3.17. The 60 MHz proton spectra of isotactic polystyrene; (a) is the aromatic and (d) the backbone proton spectrum; (a) is observed in tetrachloroethylene at 135° and (d) in o-dichlorobenzene at 200°. Spectrum (b) is a computer simulation with a linewidth of 1.0 Hz; (e) is a computer simulation with a linewidth of 3.0 Hz; (c) and (f) are the corresponding "stick" spectra (chemical shift scale in ppm from TMS).

on the right, those of the β—protons. The spectral match of the latter can be achieved only by assuming a chemical shift difference $\delta_A - \delta_B$ of 0.059 ppm, well outside experimental error at this observing frequency, and so basic principles are preserved:

The matching J couplings are $J_{AB} = -14.5$; $J_{AC} = 7.25$ Hz; $J_{BC} = 6.25$ Hz.

The interpretation of the proton spectra of atactic polystyrene (prepared with free radical or cationic initiators) is considerably less satisfactory. Some spectral simplification is achieved by deuterium substitution of the β−carbon, thus greatly reducing the effect of the vicinal (and geminal) couplings (see Section 3.5.2). Brownstein and co-workers (1961) concluded from observation of the α−proton spectrum of poly(β,β'-d_2-styrene) that the syndiotactic triad resonance predominates. Bovey *et al.* (1965) supported this interpretation. Segré *et al.* (1969) reported the spectrum of poly(d_7-styrene)

$$\cdot\cdot-CH-CD_2-\cdot\cdot$$

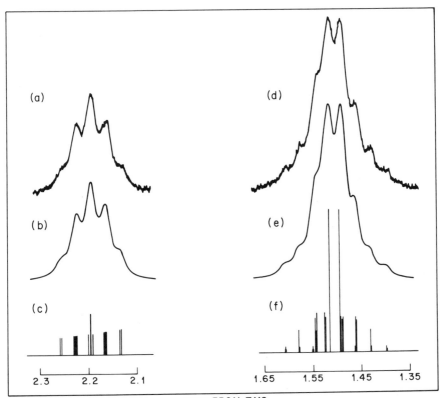

Fig. 3.18. The 220 MHz main-chain proton spectra of isotactic polystyrene, 2% in *o*-dichlorobenzene at 130°. (a,d) Experimental spectra of α− and β−protons, respectively; (b,e) calculated spectra with a linewidth of 5.5 Hz; (c,f) "stick" spectra corresponding to (b) and (e) (chemical shift scale in ppm from TMS).

and of an 80:20 copolymer (molar ratio) of d_8-styrene and d_7-styrene. Irradiation of the ^{2}H nuclei in these polymers produced at best only a slight improvement in resolution by abolishing the ^{2}H–^{1}H couplings. The spectra are shown in Fig. 3.19. It is noteworthy that even a nearly

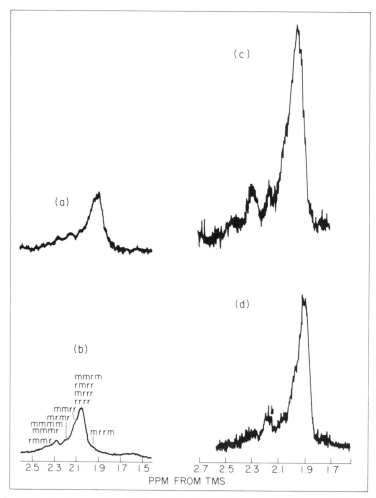

Fig. 3.19. Proton spectra of "atactic" poly(β,β'-d_2-styrene): (a) at 60 MHz in o-dichlorobenzene at 200° with ^{2}H irradiation (Bovey *et al.*, 1965); (b) at 220 MHz in o-dichlorobenzene at 150° without ^{2}H irradiation (Heatley and Bovey, 1968b); (c) 100 MHz spectrum of poly(β,β'-d_2-styrene) with perdeutero ring at 160° with ^{2}H irradiation (Segré *et al.*, 1969); (d) 100 MHz spectrum of an 80:20 copolymer (molar ratio) of poly(perdeutero styrene) and poly(β,β'-d_2-styrene) with perdeutero ring, same conditions as (c) (Segré *et al.*, 1969). Chemical shift scale in ppm from TMS.

four-fold increase in observing frequency produces little increase in apparent resolution [compare (a) with (b)] and that perdeuteration of the ring [(c) and (d)] produces at best only marginal improvement. The small chemical shift differences between pentad and longer "odd-ad" sequences are not clearly resolved and broaden the resonances so that no additional structural information can be obtained. Such behavior depends upon the relative magnitudes of the chemical shift influences of neighboring asymmetric centers and how rapidly they are attenuated with the number of intervening bonds, and these in turn depend upon the chain conformation (Chapter 7). F. Heatley and F. A. Bovey (unpublished observations, 1969) tentatively assigned the pentad chemical shifts as shown in Fig. 3.19b. We shall consider this question further when we discuss ^{13}C spectroscopy (Section 3.6). Shepherd and co-workers (1979) have reported conditions for the random epimerization (i.e., inversion) of the $\alpha-$carbons in isotactic polystyrene. They employed potassium *tert*-butoxide in hexamethylphosphoramide at 100°. The process of epimerization was followed by observation of the 300 MHz proton spectrum. At equilibrium, the spectrum closely resembled that of free radical polystyrene. (See also Section 3.6.)

3.5.4 Polypropylene

Polypropylene, in its isotactic form, is one of the major technological results of the revolution in polymer chemistry brought about by the Ziegler—Natta coordination catalysts. It is a fairly high melting (m.p. 165°) thermoplastic used in large volume in extruded and molded products. The proton spectrum is complex because the methyl and $\alpha-$protons are strongly coupled. Woodbrey (1968) has reviewed the extensive earlier literature. In Fig. 3.20 are shown the 220 MHz proton spectra ꞏ of isotactic (a) and syndiotactic (b) polypropylenes (Ferguson, 1967a,b). The isotactic polymer is prepared with initiators of the $TiCl_4$—aluminum alkyl class (Natta *et al.*, 1961). Such catalysts are heterogeneous; the $TiCl_3$ crystal surface (formed by reduction) is the active catalytic site (Bowden, 1979); it simultaneously produces chains of stereoregularity varying from nearly atactic to at least 98% isotactic. The latter is the least soluble in hydrocarbon solvents and is obtained by removal of the more soluble fractions by heptane extraction.

Syndiotactic polypropylenes are prepared with soluble coordination catalysts such as VCl_4—anisole—$Al(C_2H_5)_2Cl$ in hydrocarbons at low temperatures (Lombardi *et al.*, 1967; Zambelli and Segré, 1968; Zambelli and Tosi, 1974). They generally appear to be of somewhat

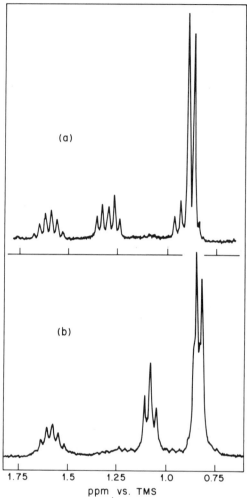

Fig. 3.20. The 220 MHz proton spectra of (a) isotactic and (b) syndiotactic polypropylenes, observed in *o*-dichlorobenzene at 165°.

lower degree of stereoregularity than can be attained with isotactic-specific ("isospecific") catalysts (above references and Zambelli *et al.,* 1973).

It will be noted in Fig. 3.20 that the β—protons of the syndiotactic polymer [spectrum (b)] appear as an apparent triplet corresponding to a single chemical shift at 1.03 ppm, whereas in the isotactic polymer [spectrum (a)] they appear as widely spaced multiplets at 1.27 and 0.87 ppm. Heatley and Zambelli (1969) employed stereospecifically deuterated monomers to demonstrate that these resonances correspond to *anti* and *syn* methylene protons, in terms of the planar zigzag conformation:

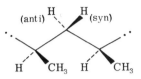

The *syn* proton is thus shielded to the same degree as the methyl protons, which appear as a doublet at 0.86 ppm. The $\alpha-$protons are relatively insensitive to stereochemical configuration, appearing at 1.57 ppm in the isotactic polymer and 1.54 ppm in the syndiotactic polymer. In terms of the spin models on p. 64, the following couplings yield accurate simulations of these spectra (Heatley *et al.*, 1969):

Isotactic : J_{AB} (geminal) : -13.5 Hz

J_{AC} $(A = H_{syn})$:   6.0 Hz

J_{BC} $(B = H_{anti})$:   7.0 Hz

$(J_{CH_3\text{-}H_C}$:   6.5 Hz)

Syndiotactic : $J_{AA'}$ (geminal) : -13.5 Hz

$J_{AB}, J_{AB'}$:   4.8, 8.3 Hz

$(J_{CH_3\text{-}H_B}$:   6.5 Hz)

The 220 MHz proton spectrum of atactic polypropylene, shown at two levels of spectrometer gain in Fig. 3.21 (Zambelli and Segré, 1968; Heatley and Zambelli, 1969), resembles those of other atactic polymers in giving an impression of degraded resolution because of the overlapping of many slightly differing chemical shifts corresponding to triad and tetrad sequences. Complete tetrad assignments have been reported by Heatley and Zambelli (1969), but some questions have been raised concerning certain of these (Stehling and Knox, 1975; Flory and Fujiwara, 1969). In particular, it appears that in highly isotactic polypropylene, the principal configurational defect corresponds to *mrr* (and *mmr*) tetrads, rather than *mrm*, as proposed by Heatley *et al.* (1969). This question is fully discussed in Section 3.6.

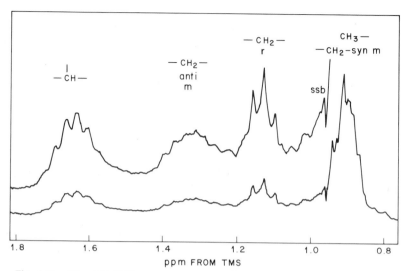

Fig. 3.21. The 220 MHz proton spectrum of atactic polypropylene observed in *o*-dichlorobenzene at 150°.

3.6 CARBON 13 SPECTROSCOPY

Carbon-13 NMR spectroscopy, despite the problems indicated in Section 2.3.4, has assumed great importance in the study of polymer structure and has in large measure superseded proton NMR, except in the study of biopolymers, where the latter still dominates. The reason for this preference lies basically in its large range of chemical shifts, over 200 ppm for common structures compared to only about $10-12$ ppm for protons. This makes it very sensitive to details of chemical structure. One can often discriminate longer configurational sequences, for example, than is possible with proton NMR. Carbon-13 NMR is particularly effective in observing hydrocarbon structures. Paraffinic protons exhibit a chemical shift range of only about 2 ppm, whereas ^{13}C nuclei are observed over a nearly 60 ppm range, and often show striking long range structural sensitivity in polymer chains.

3.6.1 Polypropylene: Propagation Errors in Coordination Catalysis

In Fig. 3.22 are shown the 25 MHz ^{13}C spectra of polypropylenes prepared with (a) an isospecific Ziegler−Natta catalyst $[TiCl_3 \cdot Al(C_2H_5)_2Cl]$; (c) a syndiospecific catalyst $[VCl_4 \cdot Al(C_2H_5)_2Cl]$; (b) is the spectrum of an atactic material. Polymer (b) was extracted with low-boiling hydrocarbon solvents from the product of the isospecific polymerization leading to (a), the major fraction of which is the isotactic polymer. As indicated, the $\beta-CH_2$, $\alpha-CH$, and methyl

carbons exhibit increasing shielding in this order. All carbon
resonances are sensitive to configuration. This is particularly clear in
spectrum (b), in which the methyl resonance is split into subpeaks
corresponding to nine of the ten possible pentad sequences
(Table 3.1). In Fig. 3.23 the methyl spectra of the atactic (a) and
isotactic (b) polymers are shown in expanded form. The isotactic

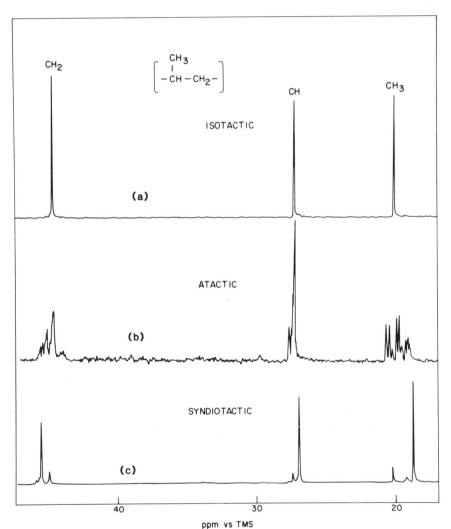

Fig. 3.22.   The 25 MHz ^{13}C spectra of three preparations of polypropylene: isotactic,
atactic, and syndiotactic, observed as 20% (w/v) solutions in 1,2,4-trichlorobenzene at
140°.

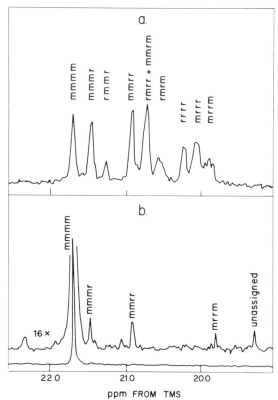

Fig. 3.23. Expanded 25 MHz methyl ^{13}C spectra of (a) atactic and (b) isotactic polypropylenes shown in Fig. 3.22.

methyl spectrum is shown at two gains. This material is 98% isotactic; the precise identification of these resonances, particularly the syndiotactic "error" resonances, is an important question with regard to the mechanism of the coordination catalyst.

We may imagine two ways in which a configurational error might be introduced into an isotactic chain. The first might be called a "steric" propagation error:

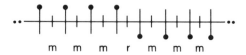

Here it is implied that the principal regulating force is the chain itself, possibly under the influence of a coordinating counterion. When an r

propagation step occurs, i.e., an "error", the correcting influences provide a new *m* configuration beyond the error, and chain propagation then proceeds as before. The associated pentad sequences are *mmmr* and *mmrm* in a 1:1 ratio. A second type of propagation error may be termed a "template" error:

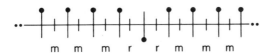

This model implies that the propagation is under the control of an asymmetric, templatelike catalyst site that is sensitive to the symmetry of the growing chain and corrects the occasional *r* "error" by restoring the absolute relationship of chain to site. The growing chain end is pictured as lying on this template, and so this mode of correction appears logical. It generates the pentads *mmmr*, *mmrr*, and *mrrm* in a 2:2:1 ratio.

Two of the peak assignments in the spectrum of the atactic polymer in Fig. 3.23 can be made by comparison to the spectra of the stereoregular polymers in Fig. 3.22. Thus the methyl resonance at lowest field (21.8 ppm) may be confidently assigned to *mmmm* since it coincides with that of the isotactic polymer; similarly, the resonance at 20.3 ppm may be assigned to *rrrr* since it coincides with the methyl resonance of the syndiotactic polymer. The other assignments shown, particularly in the heterotactic region, cannot be established in this way. Recognizing this uncertainty, Zambelli *et al.* (1975) synthesized two groups of isomeric hydrocarbons that model the pentad stereochemical sequences of polypropylene and for which the ^{13}C assignments can be made without ambiguity. These compounds are 3,5,7,9,11,13,15-heptamethylheptadecanes prepared by two routes in such a manner as to give the mixtures of diastereoisomers shown schematically in Fischer projection below as compound A and compound B. The central methyl group carbons (on C-9) are 93% enriched in ^{13}C, and only their spectra will concern us.

The synthesis produced carbons 7, 9, and 11 with approximately equal proportions of the *R* and *S* chiralities; the projections therefore represent the methyl groups on these carbons with open *circles* O in both positions. Figure 3.24 shows the four pentad environments for the methyl carbon on C-9 provided by compound A. In Fig. 3.25 the six pentad environments provided by the less symmetrical

compound B are indicated. As shown, compound B is actually
composed of eight diastereoisomers; two pairs of them, connected by

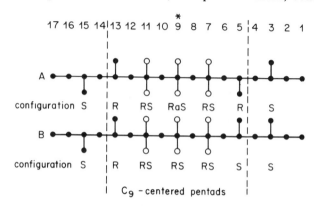

dashed lines, correspond to identical pentad sequences but different
heptad sequences, making 12 of the latter in all for both compounds
out of a possible 36 for the atactic polymer itself.

In Fig. 3.26 are shown the 25 MHz spectra of the central methyl
carbons of compound A and a mixture of compound A and
compound B. The latter duplicates very closely the resonance
positions in the atactic polymer spectrum of Fig. 3.23. If one assumes
that the methyl pentad stereochemical shifts are grouped according to
the central triad sequences *mm*, *mr*, and *rr* with increasing shielding in
this order (clearly valid) and if one further assumes that in highly
isotactic polypropylene the resonance appearing at 21.01 ppm is very
unlikely to be *mrmr*, then from the comparison of Fig. 3.26 with

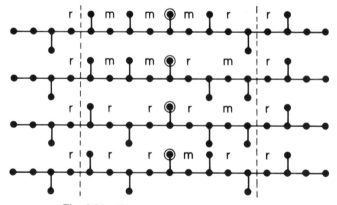

Fig. 3.24.   Diastereoisomers of compound A.

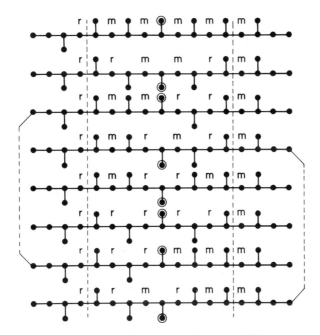

Fig. 3.25.   Diastereoisomers of compound B.

Fig. 3.23 in the light of the known isomers present in A and B, one can make the assignments shown in Fig. 3.23a. Comparing these to the residual syndiotactic resonances observed in the spectrum of the

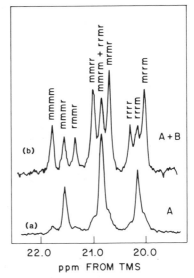

Fig. 3.26.   The 22.6 MHz ^{13}C spectra of the 9—methyl carbon (93% enriched in ^{13}C) of (a) compound A and (b) a 1:1 mixture of compound A and compound B (see text).

isotactic polymer (Fig. 3.23b) one may conclude that they correspond to *mmmr*, *mmrr*, and *mrrm* in approximately a 2:2:1 ratio as required by the "template" model. The *r* sequences are thus generated in pairs. We shall not enter here into the details of the mechanism of action of coordination catalysts, but it is generally accepted that the polymer chain grows in a hair-like fashion from the surface of the $TiCl_3$ crystals [generated by the reduction of $TiCl_4$ by the aluminum alkyl] and that a template mechanism in fact prevails. [See Bowden (1979) for further discussion.]

We shall return to a more detailed consideration of the ^{13}C spectroscopy of polypropylene in Chapters 7 and 8, where we discuss the conformational dependence of carbon chemical shifts.

3.6.2 Poly(vinyl chloride) and Poly(vinyl bromide)

We have seen (Section 3.5.2) that the proton spectrum of poly(vinyl chloride) is complex and laborious to analyze. The ^{13}C spectrum is much more readily interpreted and, like other polymer carbon spectra, does not require elaborate computer simulation. The problem of peak assignment remains, of course, but Bernoullian intensities can be assumed, since poly(vinyl chloride) is almost always prepared with free radical initiators.

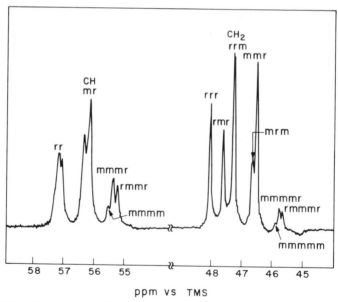

Fig. 3.27. The 25 MHz ^{13}C spectrum of poly(vinyl chloride), observed in 1,2,4-trichlorobenzene at 120° (Tonelli *et al.*, 1979).

The ^{13}C spectrum of poly(vinyl chloride) has been reported in some detail by Carman *et al.* (1971a,b; Carman, 1973), by Nishioka *et al.* (Inoue *et al.*, 1971; Ando *et al.*, 1976), by Tonelli *et al.* (1979), and by Cais and Brown (1980).

Figure 3.27 shows the 25 MHz spectrum of a typical poly(vinyl chloride). The $\alpha-$CH triad and pentad and $\beta-$CH$_2$ tetrad assignments are those proposed by Carman (1973); the $\beta-$CH$_2$ hexad assignments are those of Tonelli *et al.* (1979). The intensities of the tetrad resonances correspond to a P_m of 0.45, in approximate agreement with the value obtained from the proton spectra (p. 68). The $\alpha-$carbon chemical shifts are insensitive to solvent whereas the $\beta-$carbon chemical shifts are markedly affected. We shall consider these questions further in Chapter 7.

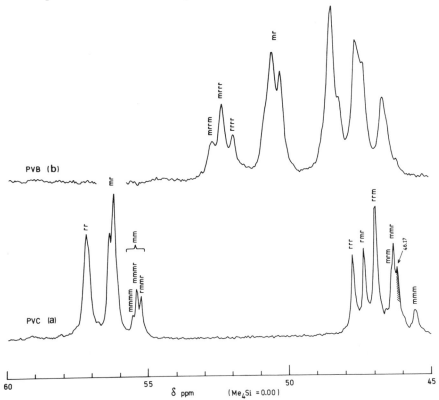

Fig. 3.28.   The 22.6 MHz ^{13}C NMR spectra of (a) poly(vinyl chloride) and (b) poly(vinyl bromide). Both spectra were obtained in 1,2,4-trichlorobenzene at 100°C. The hatched peak at 46.17 ppm in spectrum (a) is assigned to a CH$_2$Cl end group.

3. Stereochemical Configuration and Its Observation

The ^{13}C spectrum of poly(vinyl bromide) is somewhat less easy to interpret than that of poly(vinyl chloride), owing to overlap of $\alpha-$ and $\beta-$carbon resonances. In Fig. 3.28 the spectra of the two polymers are compared under the same conditions. It is evident that the marked difference between them arises mainly from the much greater shielding (by about 6 ppm) of the $\alpha-$carbon in poly(vinyl bromide). The

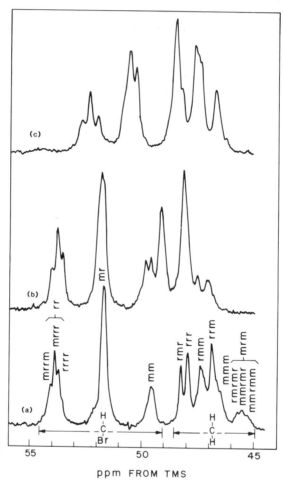

Fig. 3.29. The 22.6 MHz spectra of poly(vinyl bromide): (a) prepared at 50° in n-butyraldehyde solvent (see text), observed in carbon disulfide-acetone (1:1 by vol.) at 40°; (b) prepared at 0° (^{60}Co $\gamma-$irradiation) in bulk, observed in thietane at 80°; (c) prepared at 55° in bulk (bisazoisobutyronitrile initiator), observed in 1,2,4-trichlorobenzene at 100° (Cais and Brown, 1980).

upfield member of the α—carbon "triplet" is masked by β—carbon resonances. The latter, as in poly(vinyl chloride), are highly sensitive to solvent. This is illustrated in Fig. 3.29, which compares the spectra in (a) 1:1 (by volume) carbon disulfide:acetone; (b) thietane (trimethylene sulfide) at 80°; and (c) 1,2,4-trichlorobenzene at 100°. (Relative peak intensities, but not positions, vary in these spectra because of the differing conditions of preparation.) The best resolution is observed in (a) and the assignments to stereosequences, based in part on the study of the 2,4-dibromopentane and 2,4,6-tribromoheptane model compounds, are indicated on this spectrum. Partial resolution into α—carbon pentads can be seen; these are much better separated in 1,2,4-trichlorobenzene (c). There is also in (a) a partial resolution of the β—carbon *mrm* tetrad into its three constituent hexads.

The effect of polymerization temperature on the stereochemical configuration (see Section 3.4) of poly(vinyl chloride) has been a matter of some controversy. Earlier estimates, based mainly on proton NMR, did not agree well. Bovey *et al.* (1967) deduced a value of 310 cal (*ca.* 1.30 kJ) for solution-polymerized poly-(α-d_1-vinyl chloride). Cavalli *et al.* (1970) reported 630 cal (*ca.* 2.64 kJ) for bulk prepared polymer. The effect of temperature on vinyl bromide propagation has also been studied by proton NMR (Talamini and Vidotto, 1967) but the results were inconclusive because of poor resolution of resonances. Carbon-13 NMR has been able to provide a reliable answer to this question (Cais and Brown, 1980). Cais and Brown prepared a series of poly(vinyl chlorides) and poly(vinyl bromides) by free radical polymerization at temperatures of $-78°$ to 100°, using ^{60}Co γ-irradiation for the lower temperatures. The results are summarized in the Arrhenius plots shown in Fig. 3.30, which are plots of the data according to Eq. (3.11) (note that $P_r = 1 - P_m$) From their slopes and intercepts one obtains the values for activation enthalpies and entropies shown in Table 3.3. Thus, both monomers show a very similar dependence of tacticity upon polymerization temperature; this dependence is considerably less than that of methyl methacrylate (Section 3.4).

Another question about which there has been some disagreement is the dependence of tacticity on the solvent employed in polymerization. (Commonly, no solvent is used.) Rosen *et al.* (1961) reported that vinyl chloride polymerized in *n*-butyraldehyde (and certain other aldehydes) was more syndiotactic than the polymer prepared in bulk at the same temperature. This claim was based mainly on the more

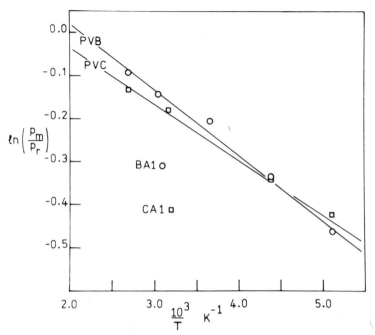

Fig. 3.30. Plot to determine the differences in activation enthalpies and entropies for meso versus racemic placement in poly(vinyl bromide) (O) and poly(vinyl chloride) (□). $\Delta H_m^{\ddagger} - \Delta H_r^{\ddagger}$ is obtained from the slope and $\Delta S_m^{\ddagger} - \Delta S_r^{\ddagger}$ from the intercept. The points corresponding to the aldehyde-modified polymers are also shown (Cais and Brown, 1980).

crystallizable nature of the material prepared in n-butyraldehyde and was denied by Bovey and Tiers (1962) because the proton spectroscopy of the time was unable to detect any difference. With the more sensitive and precise observations possible by ^{13}C NMR, Cais and Brown (1980) were able to demonstrate that in fact n-butyraldehyde does have a stereoregulating effect on both vinyl chloride and vinyl

TABLE 3.3

Temperature Dependence of Tacticity of Poly(Vinyl Chloride) and Poly(Vinyl Bromide)

monomer	$\Delta H_m^{\ddagger} - \Delta H_r^{\ddagger}$, cal (kJ)-mole^{-1}	$\Delta S_m^{\ddagger} - \Delta S_r^{\ddagger}$, cal (J)-mole^{-1}-deg^{-1}
Vinyl chloride	260 (1.09)	0.45 (1.90)
Vinyl bromide	303 (1.27)	0.65 (2.72)

bromide propagation, both yielding markedly more syndiotactic products. This is shown by the points marked CAl for vinyl chloride and BAl for vinyl bromide in Fig. 3.30. These points correspond to P_m values of 0.40 (at 40°) for vinyl chloride and 0.42 (at 50°) for vinyl bromide. The effect for vinyl chloride is more marked and is equivalent to decreasing the polymerization temperature by about 100°. It is evident that the aldehyde complexes with the growing radical chain end in some manner and "cools" it, although the nature of this interaction is unknown. The aldehyde also acts as an effective chain transfer agent (Rosen *et al.*, 1961):

$$-CH_2CHCl\cdot + RCHO \longrightarrow -CH_2CH_2Cl + RCO\cdot$$

$$RCO\cdot + CH_2{=}CHCl \longrightarrow ROCH_2CHCl\cdot \text{ etc.}$$

These end-groups could be detected in the spectra of Cais and Brown and from them number average degrees of polymerization of only 40 and 22 were obtained for poly(vinyl chloride) and poly(vinyl bromide), respectively. It is probable that these low molecular weights are in part responsible for the higher crystallizability of these materials (Böckman, 1965).

3.6.3 Polystyrene

We have seen (Section 3.5.3) that proton NMR has not satisfactorily resolved the question of the determination of the structure of stereoirregular polystyrene. Carbon-13 NMR provides somewhat more detailed information, but without as yet solving this problem. Johnson *et al.* (1970) reported that the C_1 carbon was

sensitive to stereochemical configuration and concluded that the free radical polymer was predominantly syndiotactic because most of its C_1 resonance was upfield from that of the isotactic polymer. In Fig. 3.31 is shown the 25 MHz C_1 spectrum obtained by R. E. Cais (private communication, 1976). An isotactic polymer gives a resonance in the position indicated by the dashed line. A number of authors have attempted more specific assignments of the C_1 resonances. Inoue *et al.*

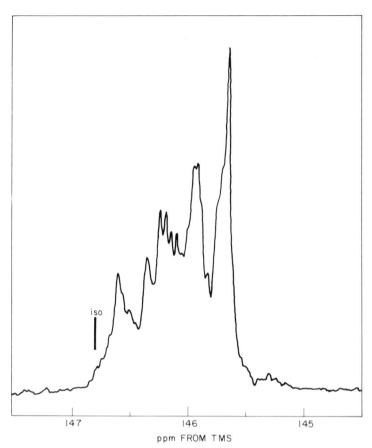

Fig. 3.31. The 25 MHz C_1 aromatic carbon spectrum in free radical polystyrene, compared to chemical shift position for C_1 in isotactic polystyrene, vertical marker. Observed as 20% (w/v) solution in 1,2,4-trichlorobenzene at 130° (R. E. Cais, private communication, 1976).

(1972) and Matsuzaki *et al.* (1972, 1975) proposed assignments consistent with the conclusion that free radical polystyrenes have a syndiotactic bias. Randall (1975, 1977) concluded from the β−carbon spectrum that it was predominantly isotactic, in contrast to all other known free radical vinyl polymers. The epimerization study by Shepherd *et al.* (1979), alluded to at the end of Section 3.5.3, was extended to the study of the β−carbon spectrum (but not the C_1 spectrum) by Chen *et al.* (1980). Again the product epimerized to equilibrium closely resembled the free radical polymer. The problem of assignment of tetrad and hexad resonances was not conclusively solved by this study, however.

Tonelli (1979) has employed the γ−gauche model (Section 7.8) to predict the β−carbon chemical shifts in styrene oligomers and has shown that the effects of stereochemical configuration are very long range and that only the central β−carbon in 2,4,6,8,10,12-hexaphenyltridecane provides a valid model for polystyrene. This theoretical approach predicts (A. E. Tonelli, private communication, 1982) that the *mmmm* pentad should appear at lowest field in the C_1 carbon spectrum, as observed (Fig. 3.31).

3.7 FLUORINE SPECTROSCOPY

Like ^{13}C, ^{19}F exhibits a very wide range of chemical shifts ($\sim$400 ppm) and shows a corresponding sensitivity to details of polymer microstructure. It also has high observing sensitivity and is therefore very useful where applicable. In this section, two examples of the application of ^{19}F NMR to the determination of stereochemical configuration are given. Further discussion is given in Section 6.2.3.

3.7.1 Polytrifluorochloroethylene

This polymer, often known by its trade name "Kel-F" (after the M. W. Kellogg Co., which commercialized it in the United States), is similar to polytetrafluoroethylene (Teflon) in resistance to chemical action and solvents, although not quite so impervious. It is highly crystalline with a melting point of 218°.

Naylor and Lasoski (1960) examined the ^{19}F spectrum of polytrifluorochloroethylene at low resolution and were able to resolve only one Cl−C−^{19}F resonance. Tiers and Bovey (1963) examined polymers of relatively low molecular weight prepared in carbon tetrachloride by free radical initiation over a 110° temperature range and compared their 40 MHz ^{19}F spectra to those of the model compounds meso and racemic $CF_2Cl \cdot CFCl \cdot CF_2 \cdot CFCl \cdot CF_2Cl$, the central three carbons of which mimic well the environment in the isotactic and syndiotactic chain sequences, respectively. The model compounds were observed in mixture, the spectra being attributable to each isomer on the basis of a differentiation of chemical shifts in the central CF_2 fluorines of the meso isomer. From the correspondence of chemical shifts in the models and the polymer, it was concluded that the polymers were about two-thirds syndiotactic (on a dyad basis) with little or no dependence on polymerization temperature.

Examination of this polymer at higher frequency by R. E. Cais (private communication, 1982) has necessitated some revision of this conclusion. In Fig. 3.32 is shown the 188 MHz ^{19}F spectrum observed in mesitylene solution at 160°. The Cl—C—^{19}F resonance is not resolved into the expected triplet even at this frequency; evidently, either *mm* or *rr* overlaps with *mr*. The CF_2 resonance, however, appears as a partially overlapping meso quartet and racemic "singlet" of approximately equal intensity. It thus may be concluded that, despite its high degree of crystallinity, the chains of polytrifluorochloro-ethylene are in fact nearly random in stereochemical configuration.

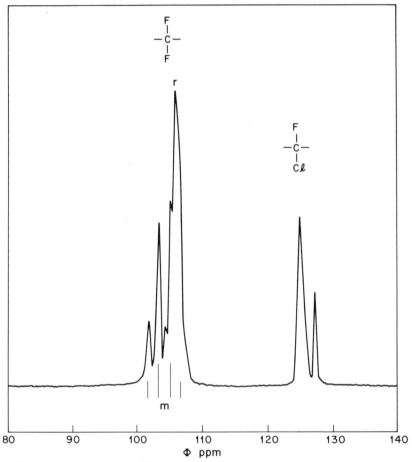

Fig. 3.32. The 188 MHz ^{19}F spectrum of polytrifluorochloroethylene observed in mesitylene solution at 160°. The chemical shift scale refers to CCl$_3$F as zero (R. E. Cais, private communication, 1982).

3.7.2 Polyfluoromethylene

When exposed to ^{60}Co γ radiation at temperatures of 0° and 37°, cis- and trans-1,2-difluoroethylene yield apparently identical polymers (Cais, 1980), assumed to have the structure

$$\left(\begin{array}{c} F \\ | \\ C \\ | \\ H \end{array}\right)_n$$

In such a chain, every carbon atom is pseudoasymmetric and only "odd-ads" have observational significance. In Fig. 3.33 is shown the 84.66 MHz ^{19}F spectrum observed in acetone solution. (The proton and ^{13}C spectra show no resolved resonances and are not helpful.) The fine structure is evidently due only to stereochemical configuration, to which ^{19}F chemical shifts are highly sensitive. The dotted spectrum in Fig. 3.33 is a computer simulation with pentad resonance positions as shown and with intensities based on the assumption of Bernoullian statistics with a P_m of 0.42. This is in conformity with the general observation that free radical polymers tend to be predominantly syndiotactic. The propagation step in 1,2-disubstituted monomers generates pseudoasymmetric centers two at a time and it is not clear that Bernoullian statistics need necessarily prevail, but any departure cannot be large. (It is possible, in fact, that P_m may be 0.58, in which case all assignments must be altered by exchanging m for r and vice versa.)

It should be noted (see Section 1.6 and the Appendix to this chapter) that in such chains, the Fischer and planar zigzag projections conflict:

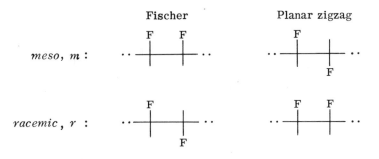

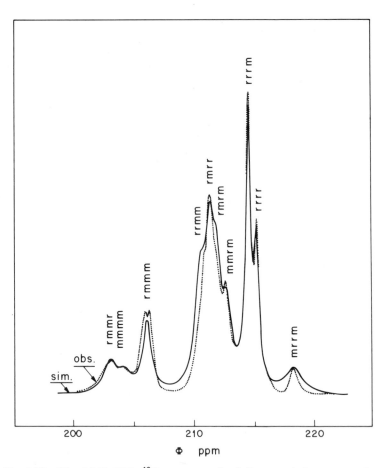

Fig. 3.33. The 84.66 MHz ^{19}F spectrum of polyfluoromethylene prepared at 0°, observed in acetone solution. The chemical shift scale refers to CCl_3F as zero. The observed spectrum is the solid line. The dotted line is a computer simulation which is further explained in the text.

APPENDIX

The distinction between Fischer and planar zigzag conventions for the representation of the stereochemistry of polymer chains, particularly vinyl polymers, should be made clear since confusion can readily occur. This is particularly important since the International Union of Pure and Applied Chemistry has adopted the Fischer convention for this purpose (Jenkins, 1981).

The Fischer scheme represents asymmetric carbons (and asymmetric centers in general) in the manner illustrated on the right below for 2,3-dichlorobutane:

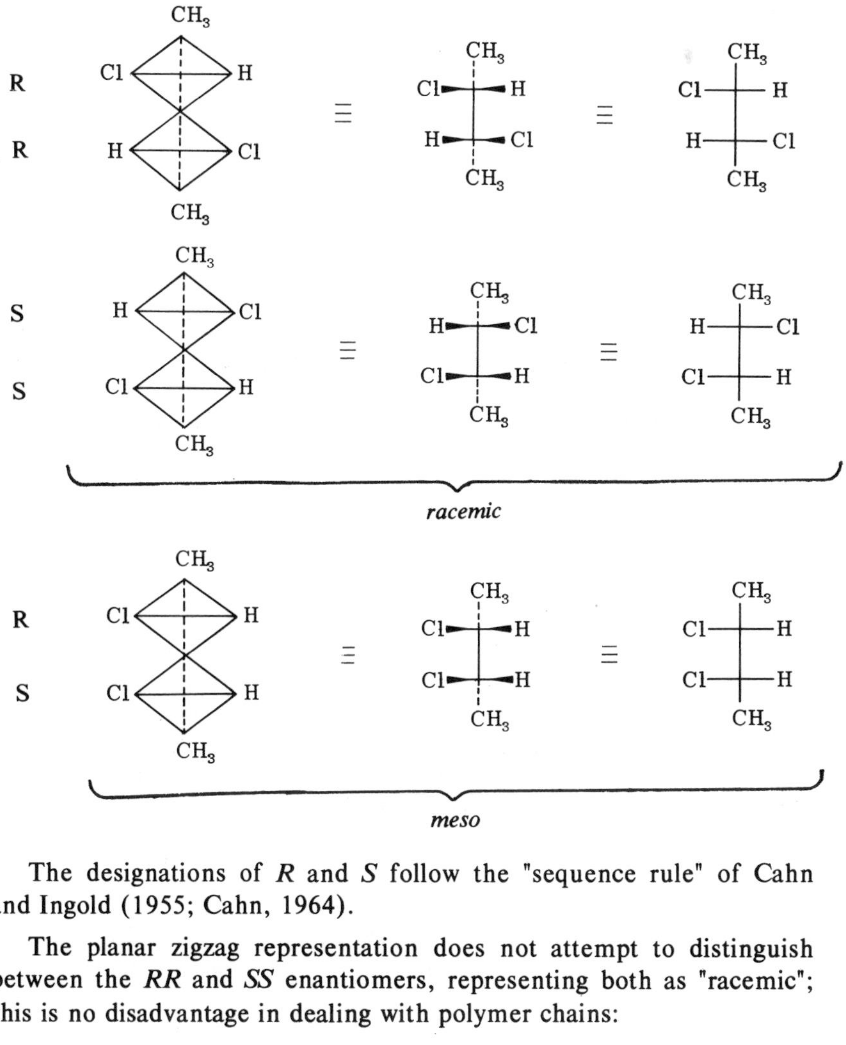

R R

$$\equiv$$

S S

racemic

R S

$$\equiv$$

meso

The designations of R and S follow the "sequence rule" of Cahn and Ingold (1955; Cahn, 1964).

The planar zigzag representation does not attempt to distinguish between the RR and SS enantiomers, representing both as "racemic"; this is no disadvantage in dealing with polymer chains:

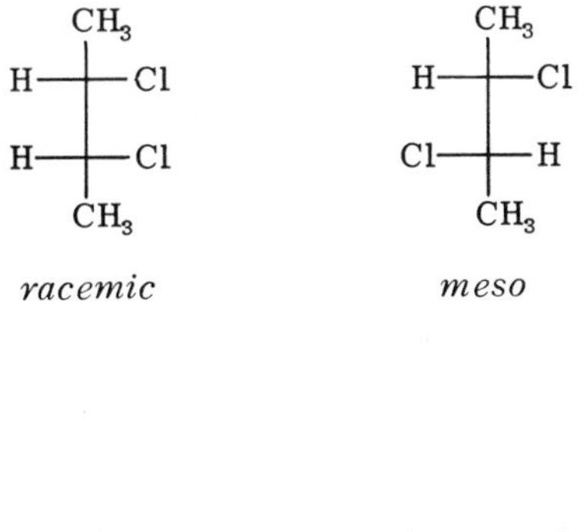

racemic   meso

It can be seen that this case presents an analogy to polyfluoromethylene, $(CHF)_n$, discussed in Section 3.7.2, in that the Fischer and planar zigzag projections differ and therefore must be identified. This will also be true where the substituents are unlike (Chapter 1, p. 6):

Fischer: (left) erythro — CH₃, A—H, B—H, CH₃; (right) threo — CH₃, A—H, H—B, CH₃

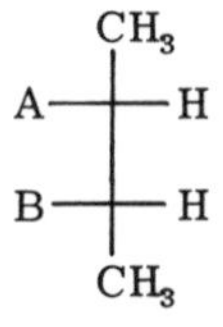

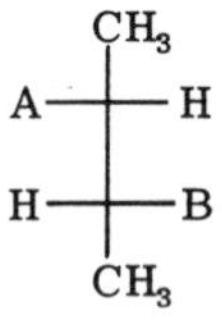

(Absolute chiralities cannot be specified as they will depend on the rank ordering of A and B in the sequence rule.)

Planar zigzag: (left) erythro — CH₃, A—H, H—B, CH₃; (right) threo — CH₃, A—H, B—H, CH₃

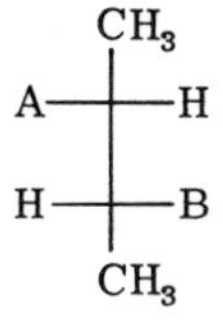

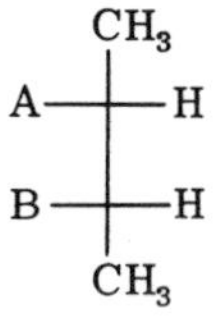

For the representation of 2,4-disubstituted pentanes, 2,4,6-trisubstituted heptanes, and α—substituted vinyl polymer chains, the Fischer and planar zigzag projections appear to coincide, although actually, with regard to the model compounds, the Fischer projection implies a knowledge of absolute chiralities. With respect to long chains (end-groups not represented) no such significance is implied for either projection. Thus, we may represent R, S, and meso 2,4-chloropentanes as

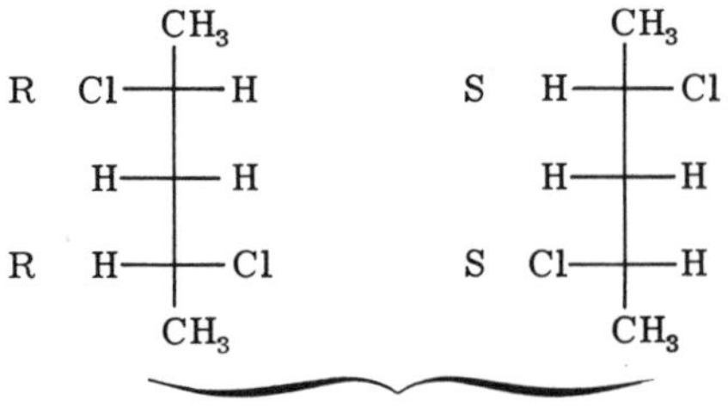

racemic 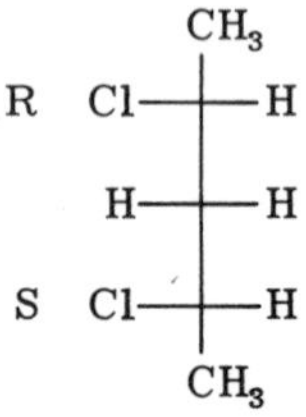   meso

For a long chain, the projection is usually rotated 90° to save space and only relative chiralities are implied (except for the somewhat artificial treatment discussed in Section 7.4):

$$\cdots \underset{\underset{H}{|}}{\overset{\overset{Cl}{|}}{}} \underset{\underset{H}{|}}{\overset{\overset{H}{|}}{}} \underset{\underset{H}{|}}{\overset{\overset{Cl}{|}}{}} \underset{\underset{H}{|}}{\overset{\overset{H}{|}}{}} \underset{\underset{Cl}{|}}{\overset{\overset{H}{|}}{}} \underset{\underset{H}{|}}{\overset{\overset{H}{|}}{}} \underset{\underset{Cl}{|}}{\overset{\overset{H}{|}}{}} \underset{\underset{H}{|}}{\overset{\overset{H}{|}}{}} \cdots$$

For chains with a three-atom repeat, such as polypropylene oxide, the Fischer and planar zigzag representations again conflict. Thus, an isotactic chain (Section 1.6) is represented in planar zigzag projection as

$$\cdots - \underset{\underset{H}{|}}{\overset{\overset{CH_3}{|}}{}} CH_2 - O - \underset{\underset{CH_3}{|}}{\overset{\overset{H}{|}}{}} CH_2 - O - \underset{\underset{H}{|}}{\overset{\overset{CH_3}{|}}{}} CH_2 - O - \cdots$$

The Fischer projection is

$$\cdots - \underset{\underset{H}{|}}{\overset{\overset{CH_3}{|}}{}} CH_2 - O - \underset{\underset{H}{|}}{\overset{\overset{CH_3}{|}}{}} CH_2 - O - \underset{\underset{H}{|}}{\overset{\overset{CH_3}{|}}{}} CH_2 - O - \cdots$$
$$\quad S \qquad\qquad S \qquad\qquad S$$

or

$$\cdots - \underset{\underset{CH_3}{|}}{\overset{\overset{H}{|}}{}} CH_2 - O - \underset{\underset{CH_3}{|}}{\overset{\overset{H}{|}}{}} CH_2 - O - \underset{\underset{CH_3}{|}}{\overset{\overset{H}{|}}{}} CH_2 - O - \cdots$$
$$\quad R \qquad\qquad R \qquad\qquad R$$

Here, the Fischer projection has the advantage of being able to represent the chiralities of carbons which are in this case true asymmetric centers. In general, however, the planar zigzag conformation is much easier to visualize or to construct with models than the Fischer conformation, which curls back on itself in a manner that is very awkward when more than five or six main chain atoms are involved.

REFERENCES

Ando, I., Kato, Y., and Nishioka, A. (1976). *Makromol. Chem.* **177**, 2759.

Böckman, O. Chr. (1965). *J. Polymer Sci., Part A* **3**, 3399.

Bovey, F. A. (1960). *J. Polymer Sci.* **46**, 59.

Bovey, F. A., and Tiers, G. V. D. (1960). *J. Polymer Sci.* **44**, 173.

Bovey, F. A., and Tiers, G. V. D. (1962). *Chem. Ind.* **1962**, 1826.

Bovey, F. A., Tiers, G. V. D., and Filipovich, G. (1959). *J. Polymer Sci.* **38**, 73.

Bovey, F. A., Anderson, E. W., Douglass, D. C., and Manson, J. A. (1963). *J. Chem. Phys.* **39**, 1199.

Bovey, F. A., Hood, F. P., Anderson, E. W., and Snyder, L. C. (1965). *J. Chem. Phys.* **42**, 3900.

Bovey, F. A., Hood, F. P., Anderson, E. W., and Kornegay, R. L. (1967). *J. Phys. Chem.* **71**, 312.

Bowden, M. (1979). *In* "Macromolecules" (F. A. Bovey and F. H. Winslow, eds.). Academic Press, New York.

Brownstein, S., Bywater, S., and Worsfold, D. J. (1961). *Makromol. Chem.* **48**, 127.

Cahn, R. S. (1964). *J. Chem. Ed.* **41**, 116.

Cahn, R. S., and Ingold, C. K. (1955). *J. Chem. Soc.* **1955**, 612.

Cais, R. E. (1980). *Macromolecules* **13**, 806.

Cais, R. E., and Brown, W. L. (1980). *Macromolecules* **13**, 801.

Carman, C. J. (1973). *Macromolecules* **6**, 725.

Carman, C. J., Tarpley, A. R., Jr., and Goldstein, J. H. (1971a). *J. Am. Chem. Soc.* **93**, 2864.

Carman, C. J., Tarpley, A. R., Jr., and Goldstein, J. H. (1971b). *Macromolecules* **4**, 445.

Cavalli, L., Borsini, G. C., Carraro, G., and Confalonieri, G. (1970). *J. Polymer Sci., Part A-1* **8**, 801.

Chen, T. K., Gerken, T. A., and Harwood, H. J. (1980). *Polymer Bull.* **2**, 37.

Chûjô, R., Satoh, S., Ozeki, T., and Nagai, E. (1962). *J. Polymer Sci.* **61**, S12.

Ferguson, R. C. (1967a). *Polymer Preprints* **8**(2), 1026.

Ferguson, R. C. (1967b). *Trans. N.Y. Acad. Sci.* **29**, 495.

Flory, P. J., and Baldeschwieler, J. D. (1966). *J. Am. Chem. Soc.* **88**, 2873.

Flory, P. J., and Fujiwara, Y. (1969). *Macromolecules* **2**, 327.

Fox, T. G., and Schnecko, H. W. (1963). *Polymer* **3**, 575.

Frisch, H. L., Mallows, C. L., and Bovey, F. A. (1966). *J. Chem. Phys.* **45**, 1565.

Frisch, H. L., Mallows, C. L., Heatley, F., and Bovey, F. A. (1968). *Macromolecules* **1**, 533.

Heatley, F., and Bovey, F. A. (1968a). *Macromolecules* **1**, 303.

Heatley, F., and Bovey, F. A. (1968b). *Macromolecules* **1**, 301.

Heatley, F., and Bovey, F. A. (1969). *Macromolecules* **2**, 241.

Heatley, F., and Zambelli, A. (1969). *Macromolecules* **2**, 618.

Heatley, F., Salovey, R., and Bovey, F. A. (1969). *Macromolecules* **2**, 619.

Inoue, Y., Ando, I., and Nishioka, A. (1971). *Polymer J.* **3**, 246.

Inoue, Y., Nishioka, A., and Chûjô, R. (1972). *Makromol. Chem.* **156**, 207.

Jenkins, A. D. (1981). *Pure Appl. Chem.* **53** 733.

Johnsen, U. (1961). *J. Polymer Sci.* **54**, S 6.

Johnson, C. E., Jr., and Bovey, F. A. (1958). *J. Chem. Phys.* **29**, 1012.

Johnson, L. F., Heatley, F., and Bovey, F. A. (1970). *Macromolecules* **3**, 175.

Krimm, S. (1960). *Fortschr. Hochpolym.-Forsch.*, **2**, 51.

Krimm, S., Berens, A. R., Folt, V. L., and Shipman, J. J. (1958). *Chem. Ind.* **1958**, 1512.

Krimm, S., Berens, A. R., Folt, V. L., and Shipman, J. J. (1959). *Chem. Ind.* **1959**, 433.

Lombardi, E., Segré, A. L., Zambelli, A., and Marinangeli, A. (1967). *J. Polymer Sci., Part C* **16**, 2539.

Matsuzaki, K., Uryu, T., Osada, K., and Kawamura, T. (1972). *Macromolecules* **5**, 816.

Matsuzaki, K., Uryu, T., Seki, T., Osada, K., and Kawamura, T. (1975). *Makromol. Chem.* **176**, 3051.

Natta, G., and Corradini, P. (1956). *J. Polymer Sci.* **20**, 251.

Natta, G., Pasquon, I., Zambelli, A., and Gatti, G. (1961). *J. Polymer Sci.* **51**, 387.

Naylor, R. E., and Lasoski, Jr., S. W. (1960). *J. Polymer Sci.* **44**, 1.

Odajima, A. (1959). *J. Phys. Soc. Japan* **14**, 777.

Otsu, T., Yamada, B., and Imoto, M. (1966). *J. Macromol. Chem.* **1**, 61.

Randall, J. C. (1975). *J. Polymer Sci. Polymer Phys. Ed.* **13**, 889.

Randall, J. C. (1977). "Polymer Sequence Determination," pp. 87-92, 116-119. Academic Press, New York.

Rosen, I., Burleigh, P. H., and Gillespie, J. F. (1961). *J. Polymer Sci.* **54**, 31.

Saunders, M., and Wishnia, A. (1958). *Ann. N.Y. Acad. Sci.* **70**, 870.

Schuerch, C., Fowells, W., Yamada, A., Bovey, F. A., and Hood, F. P. (1964). *J. Am. Chem. Soc.* **86**, 4481.

Segré, A. L., Ferruti, P., Toja, E., and Danusso, F. (1969). *Macromolecules* **2**, 35.

Shepherd, L., Chen, T. K., and Harwood, H. J. (1979). *Polymer Bull.* **1**, 445.

Stehling, F. C., and Knox, J. R. (1975). *Macromolecules* **8**, 595.

Talamini, G., and Vidotto, G. (1967). *Makromol. Chem.* **100**, 48.

Tiers, G. V. D., and Bovey, F. A. (1963). *J. Polymer Sci., Part A* **1**, 833.

Tincher, W. C. (1962). *J. Polymer Sci.* **62**, S 148.

Tonelli, A. E. (1979). *Macromolecules* **12**, 252.

Tonelli, A. E., Schilling, F. C., Starnes, Jr., W. H., Shepherd, L., and Plitz, I. M. (1979). *Macromolecules* **12**, 78.

van Gorkom, M., and Hall, G. E. (1968). *Quart. Rev. Chem. Soc.* **22**, 14.

Waugh, J. S., and Fessenden, R. W. (1957). *J. Am. Chem. Soc.* **79**, 846.

Waugh, J. S., and Fessenden, R. W. (1958). *J. Am. Chem. Soc.* **80**, 6697.

Woodbrey, J. C. (1968). *In* "The Stereochemistry of Macromolecules" (A. D. Ketley, ed.), Vol. 3. Marcel Dekker. New York.

Zambelli, A., and Segré, A. L. (1968). *J. Polymer Sci., Part B* **6**, 473.

Zambelli, A., and Tosi, C. (1974). *Adv. Polymer Sci.* **15**, 31.

Zambelli, A., Dorman, D. E., Brewster, A. I. R., and Bovey, F. A. (1973). *Macromolecules* **6**, 925.

Zambelli, A., Locatelli, P., Bajo, G., and Bovey, F. A. (1975). *Macromolecules* **8**, 687.

Chapter 4

GEOMETRICAL ISOMERISM IN DIENE POLYMERS

4.1 INTRODUCTION

We have seen (Section 1.4) that diene monomer units may assume many forms when enchained together in sequences in macromolecules. These forms may yield quite different physical properties and it has long been a matter of importance to be able to observe and measure them. In this chapter, we shall describe these methods and the results obtained with them. Isomerism in diene polymer chains can be readily detected by both NMR and vibrational spectroscopy.

4.2 POLYBUTADIENE

In Fig. 4.1 are shown infrared spectra in the $700-1200$ cm^{-1} out-of-plane bending region of (a) cis-1,4- (or Z-1,4-), (b) trans-1,4- (or E-1,4-) and (c) 1,2-polybutadienes, prepared with specific catalysts (G. Pasteur, private communication, 1975). The cis polymer shows a broad strong band at 740 cm^{-1}, whereas in the trans polymer the

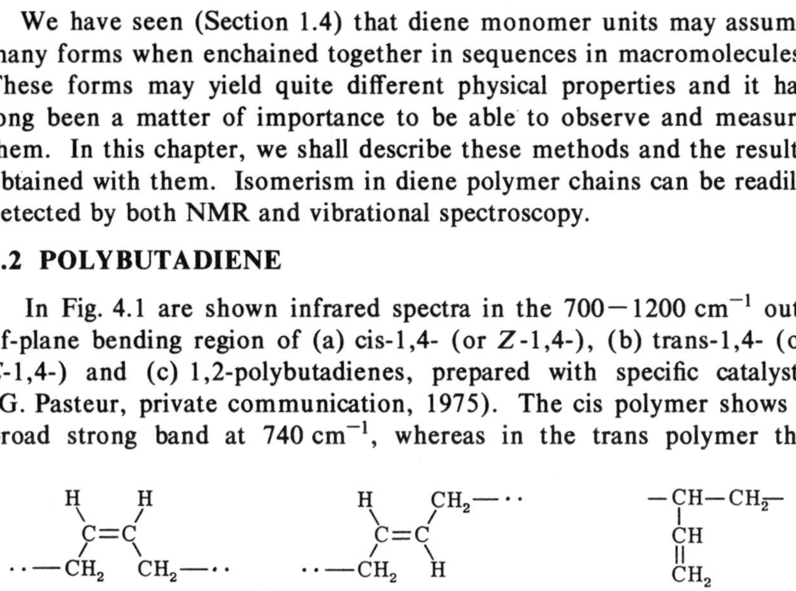

$$
\begin{array}{ccc}
\underset{\displaystyle \substack{\cdots-CH_2 \quad CH_2-\cdots}}{\overset{\displaystyle \substack{H \qquad H}}{\underset{\displaystyle}{C=C}}} &
\underset{\displaystyle \substack{\cdots-CH_2 \quad H}}{\overset{\displaystyle \substack{H \qquad CH_2-\cdots}}{\underset{\displaystyle}{C=C}}} &
\substack{-CH-CH_2- \\ | \\ CH \\ \| \\ CH_2}
\\[2em]
(a) & (b) & (c)
\end{array}
$$

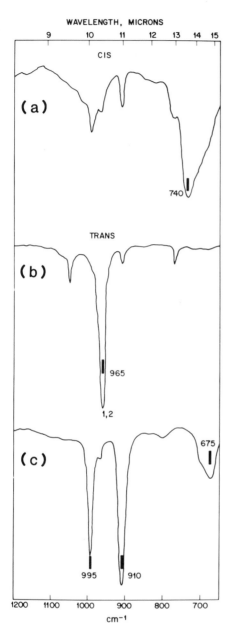

Fig. 4.1. Out-of-plane bending bands of (a) cis-1,4-polybutadiene, (b) trans-1,4-polybutadiene, and (c) 1,2-polybutadiene (G. Pasteur, private communication, 1975).

corresponding band appears at $965\ cm^{-1}$ and is narrower. 1,2-Polybutadiene shows the vinyl CH ($995\ cm^{-1}$) and vinyl CH_2 ($910\ cm^{-1}$) bending vibrations that we have already seen

(Fig. 2.9) as a minor feature of the spectrum of linear polyethylene. The spectra show that none of the three polymers is entirely regular. The trans polymer contains some 1,2 structures and the cis polymer contains both 1,2 and trans, while the 1,2 polymer exhibits a cis band at *ca.* 670 cm^{-1}. The spectra can yield quantitative analysis of the polymers by careful intercomparison even if extinction coefficients are not established.

For purely analytical purposes, infrared and Raman spectroscopy are very widely employed because they are relatively quick and easy and are adequately quantitative if proper calibration is employed. For accurate measurement of minor isomeric structures and for more sophisticated information concerning monomer sequences, NMR is indispensable. It will be the principal topic of the subsequent discussion.

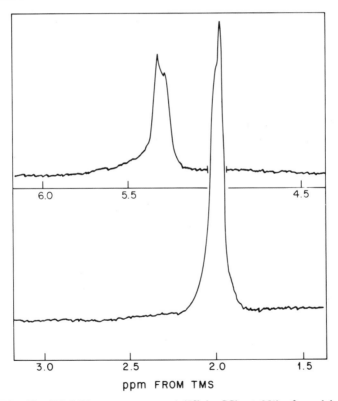

Fig. 4.2. The 220 MHz proton spectrum (5% in CCl$_4$ at 25°) of a polybutadiene containing 52% of trans-1,4 and 48% of cis-1,4 units (F. A. Bovey and A. I. Brewster, unpublished observations, 1971).

In Fig. 4.2 is shown the 220 MHz proton spectrum of a polybutadiene containing 52% of trans-1,4 and 48% of cis-1,4 units. Both the methylene resonance at 2.0 ppm and the vinylene resonance at *ca.* 5.3 ppm show distinct shoulders which evidently reflect the cis units. At any lower observing frequency the two units would be indistinguishable. At 300 MHz, the resonances are more fully resolved (Santee *et al.,* 1973a,b). On the other hand, the presence of 1,2 units can be readily distinguished. Figure 4.3 shows the 100 MHz spectrum of a polybutadiene containing about 98% 1,2-linked monomer units. The vinyl protons appear as a complex multiplet near 5 ppm; the vinyl methylene protons are centered at about 4.8 ppm and the methine proton at 5.35 ppm. These resonances are complicated by scalar coupling and by stereochemical configuration.

Figure 4.4 shows the 15.08 MHz ^{13}C spectra of (a) cis-1,4-polybutadiene and (b) trans-1,4-polybutadiene, reported in a

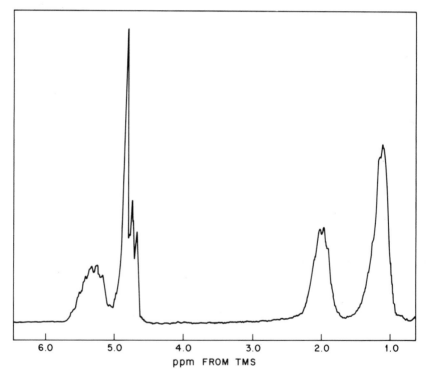

Fig. 4.3. The 100 MHz proton spectrum (5% in CS_2 at 45°) of a polybutadiene containing about 98% of 1,2-linked monomer units (F. A. Bovey and A. I. Brewster, unpublished observations, 1971).

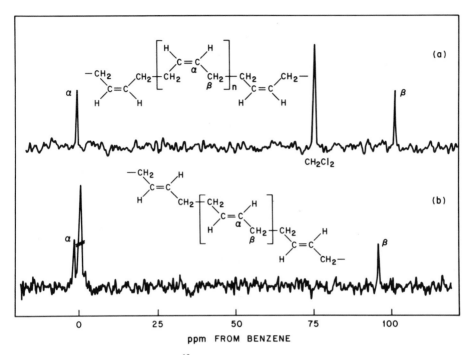

Fig. 4.4.   The 15.08 MHz ^{13}C NMR spectrum of (a) cis-1,4-polybutadiene and (b) trans-1,4-polybutadiene. Spectrum (a) was obtained at 45° using a 20% (w/v) solution in CH_2Cl_2; spectrum (b) was obtained at 45° using a benzene solution. (Duch and Grant, 1970).

pioneering study by Duch and Grant (1970). It is evident that the methylene carbons are sensitive to geometrical isomerism, being more shielded in the cis isomer by approximately 8 ppm. (This is no doubt related to *gauche* shielding observed in saturated chains and discussed in Section 7.8).

Chains of mixed structure exhibit more complex spectra because of sequence effects. (Mochel, 1972; Hatada *et al.*, 1973; Furukawa *et al.*, 1974; Conti *et al.*, 1974a,b; Elgert *et al.*, 1974; Tanaka *et al.*, 1974, 1977; Julemont *et al.*, 1974; Clague *et al.*, 1974; Walckiers and Julemont, 1981). Curiously, however, ^{13}C NMR, normally more discriminating than proton NMR, has proved misleading because of insensitivity to cis—trans sequences. The spectra are not complex enough! Mochel (1972) reported that polybutadiene prepared with a n—BuLi catalyst showed no sequences of cis- and trans-1,4 units because the CH_2 resonances showed no fine structure. He concluded that the cis and trans units occurred in runs separated by isolated

1,2 units. It was then shown (Santee *et al.,* 1973a,b) that at 300 MHz the olefinic proton resonances corresponding to cis and trans double bonds are each split into triplets. These must arise from trans—trans—trans, trans—trans—cis, cis—trans—cis, cis—cis—cis, cis—cis—trans, and trans—cis—trans sequences, and their intensities corresponded to a random (or Bernoullian) distribution. The methylene carbon resonances of the polymer showed no corresponding splitting.

In Fig. 4.5 is shown the 50 MHz ^{13}C spectrum of a free radical polybutadiene (L. W. Jelinski, private communication, 1982); the right part of the spectrum (a) shows the olefinic carbons (not assigned in detail), and the left part (b) the aliphatic carbons. The assignments, in agreement with those of earlier studies, are based mainly on the

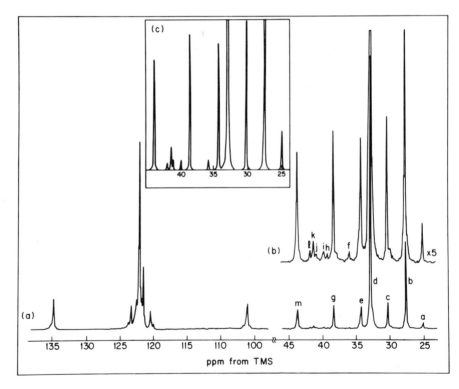

Fig. 4.5. The 50 MHz ^{13}C spectrum of free radical polybutadiene, observed at 50° in 20 wt% solution in CDCl$_3$ (L. W. Jelinski, private communication, 1982). (a) Olefinic carbon spectrum; (b) aliphatic carbon spectrum; (c) computer simulation of (b) based on a random distribution of cis, trans, and 1,2 units in the proportion 23:58:19.

chemical shift parameters of Grant and Paul (1964). The assignments in part (b) are shown in Table 4.1.

TABLE 4.1

[13]C Peak Assignments for Free Radical Polybutadiene

Peak designation[a]	Chemical shift[b]	Assignment
a	25.2	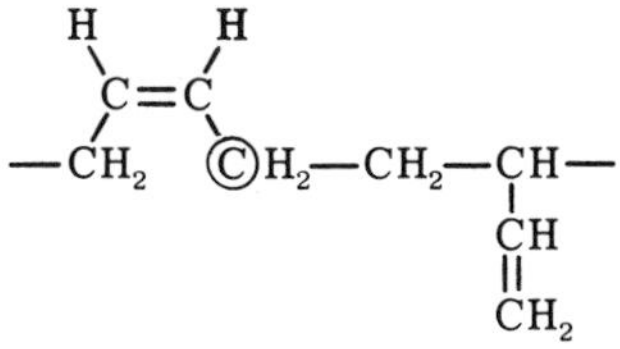
b	27.4	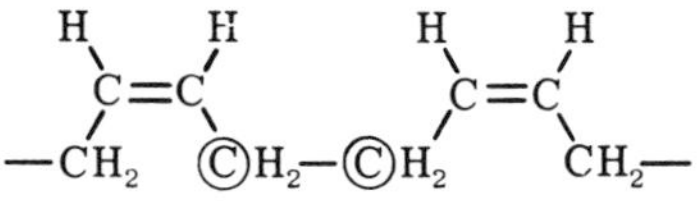
c	30.2	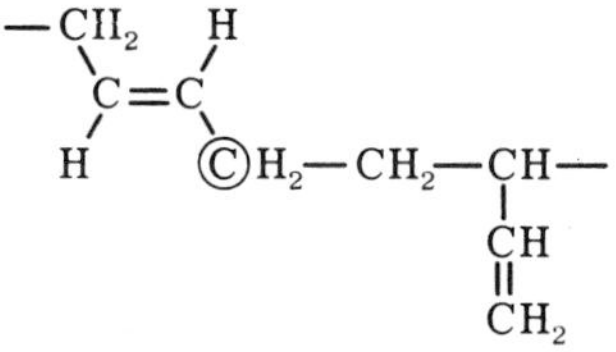
d	32.7	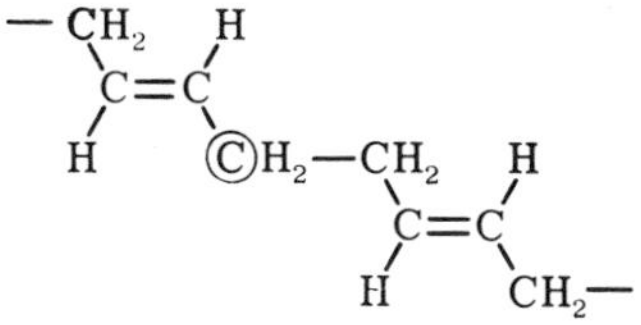

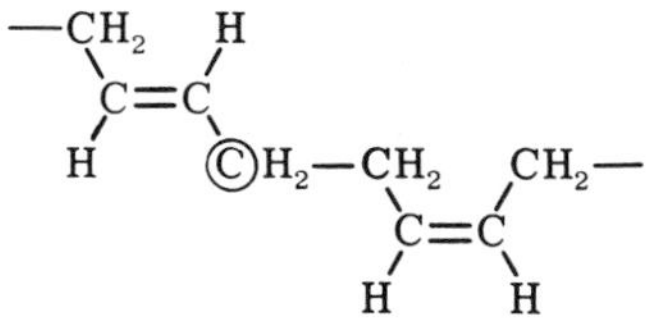

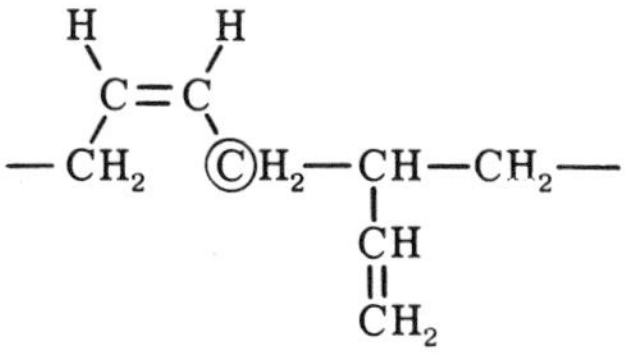

TABLE 4.1, CONTINUED

Peak designation[a]	Chemical shift[b]	Assignment

e   34.1

$$-CH_2 \quad CH_2-CH-\textcircled{C}H_2-CH_2 \quad CH_2-$$
$$\quad\backslash C=C \diagup \qquad \mid \qquad\qquad\qquad C=C$$
$$\quad H \qquad H \qquad CH \qquad\qquad\qquad H \qquad H$$
$$\qquad\qquad\qquad\qquad \| $$
$$\qquad\qquad\qquad\qquad CH_2$$

$$-CH_2 \quad H$$
$$\quad\backslash C=C$$
$$\quad H \qquad\backslash CH_2-CH-\textcircled{C}H_2-CH_2 \quad CH_2-$$
$$\qquad\qquad\qquad \mid \qquad\qquad\qquad C=C$$
$$\qquad\qquad\qquad CH \qquad\qquad\qquad H \qquad H$$
$$\qquad\qquad\qquad \|$$
$$\qquad\qquad\qquad CH_2$$

$$-CH_2 \quad H$$
$$\quad\backslash C=C$$
$$\quad H \qquad\backslash CH_2-CH-\textcircled{C}H_2-CH_2 \quad H-$$
$$\qquad\qquad\qquad \mid \qquad\qquad\qquad C=C$$
$$\qquad\qquad\qquad CH \qquad\qquad\qquad H \quad\backslash CH_2-$$
$$\qquad\qquad\qquad \|$$
$$\qquad\qquad\qquad CH_2$$

f   35.9

$$-CH-CH_2-CH-\textcircled{C}H_2-CH_2 \quad CH_2-$$
$$\quad \mid \qquad\qquad \mid \qquad\qquad\qquad C=C$$
$$\quad CH \qquad\qquad CH \qquad\qquad\qquad H \qquad H$$
$$\quad \| \qquad\qquad\quad \|$$
$$\quad CH_2 \qquad\qquad CH_2$$

$$-CH-CH_2-CH-\textcircled{C}H_2-CH_2 \quad H$$
$$\quad \mid \qquad\qquad \mid \qquad\qquad\qquad C=C$$
$$\quad CH \qquad\qquad CH \qquad\qquad\qquad H \quad\backslash CH_2-$$
$$\quad \| \qquad\qquad\quad \|$$
$$\quad CH_2 \qquad\qquad CH_2$$

g   39.0

$$-CH-CH_2-CH-\textcircled{C}H_2-CH-CH_2-$$
$$\quad \mid \qquad\qquad \mid \qquad\qquad\qquad \mid$$
$$\quad CH \qquad\qquad CH \qquad\qquad\qquad CH$$
$$\quad \| \qquad\qquad\quad \| \qquad\qquad\qquad \|$$
$$\quad CH_2 \qquad\qquad CH_2 \qquad\qquad\qquad CH_2$$

h   39.1

$$-CH_2 \quad H$$
$$\quad\backslash C=C$$
$$\quad H \qquad\textcircled{C}H_2-CH-CH_2-$$
$$\qquad\qquad\qquad \mid$$
$$\qquad\qquad\qquad CH$$
$$\qquad\qquad\qquad \|$$
$$\qquad\qquad\qquad CH_2$$

TABLE 4.1, CONTINUED

Peak designation[a]	Chemical shift[b]	Assignment

i — 40.4

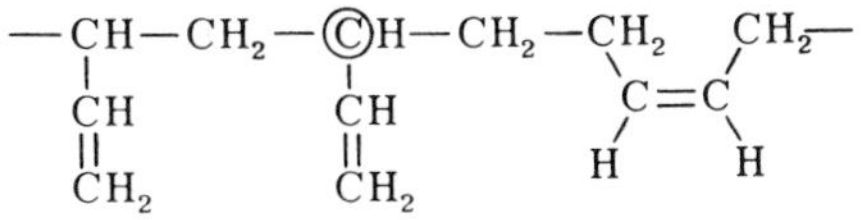

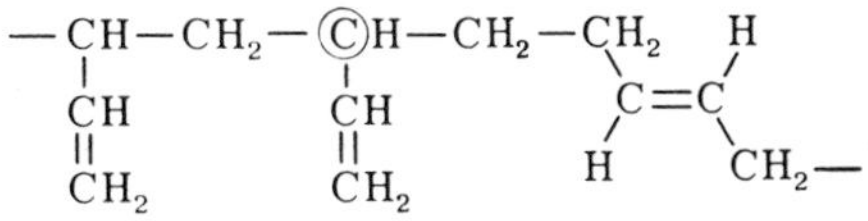

$$—CH—CH_2—\textcircled{C}H—CH_2—CH_2 \quad CH_2—$$
with CH, CH₂ branches and C=C (H, H)

$$—CH—CH_2—\textcircled{C}H—CH_2—CH_2 \quad H$$
with CH, CH₂ branches and C=C (H, CH₂—)

j — 41.2

$$—CH_2—CH—CH_2—\textcircled{C}H—CH_2 \quad CH_2—$$
with CH, CH₂ branches and C=C (H, H)

$$—CH_2—CH—CH_2—\textcircled{C}H—CH_2 \quad H$$
with CH, CH₂ branches and C=C (H, CH₂—)

k — 41.3

$$—CH_2—CH—\textcircled{C}H_2—CH—CH_2 \quad CH_2—$$
with CH, CH₂ branches and C=C (H, H)

$$—CH_2—CH—\textcircled{C}H_2—CH—CH_2 \quad H$$
with CH, CH₂ branches and C=C (H, CH₂—)

l — 41.7

$$—CH—\textcircled{C}H_2—CH—\textcircled{C}H_2—CH—CH_2—$$
with CH, CH₂ branches (three)

m — 43.6

$$—CH_2 \quad CH_2—\textcircled{C}H—CH_2—CH_2 \quad CH_2—$$
with C=C (H, H), CH, CH₂ branch, C=C (H, H)

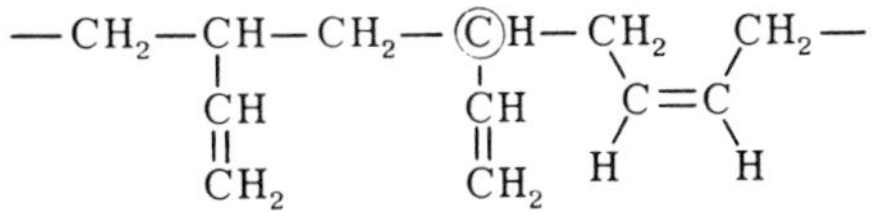

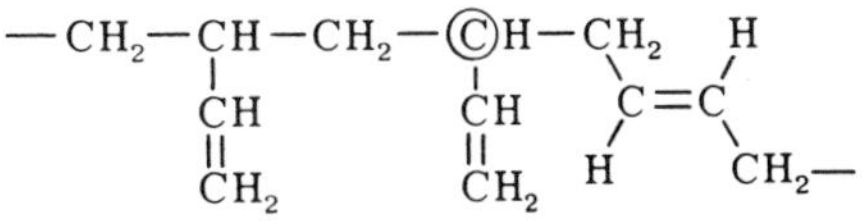

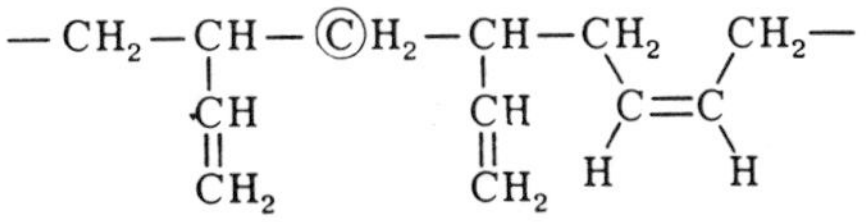

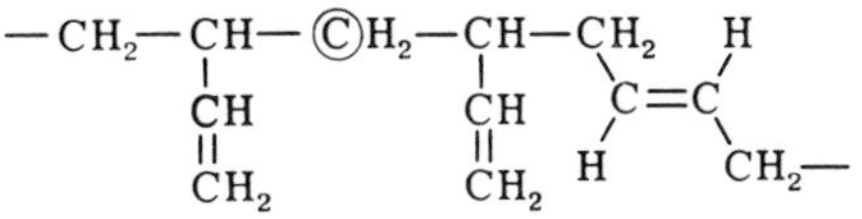

$$—CH_2 \quad CH_2—\textcircled{C}H—CH_2—CH_2 \quad H$$
with C=C (H, H), CH, CH₂ branch, C=C (H, CH₂—)

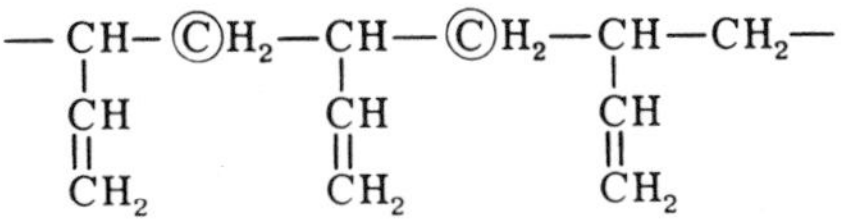

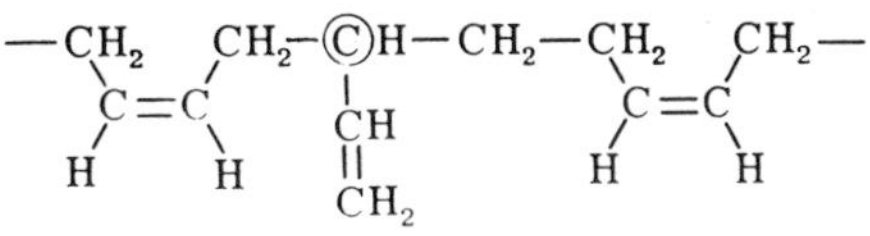

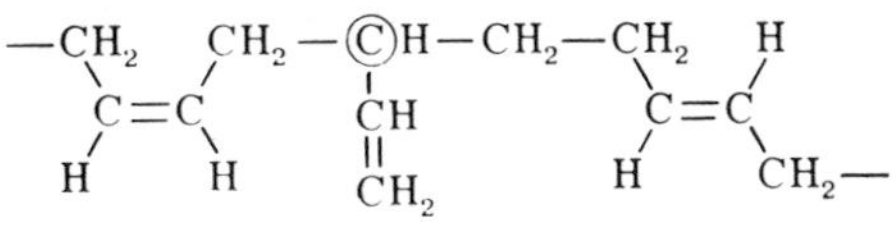

TABLE 4.1, CONTINUED

Peak designation[a]	Chemical shift[b]	Assignment
m, continued		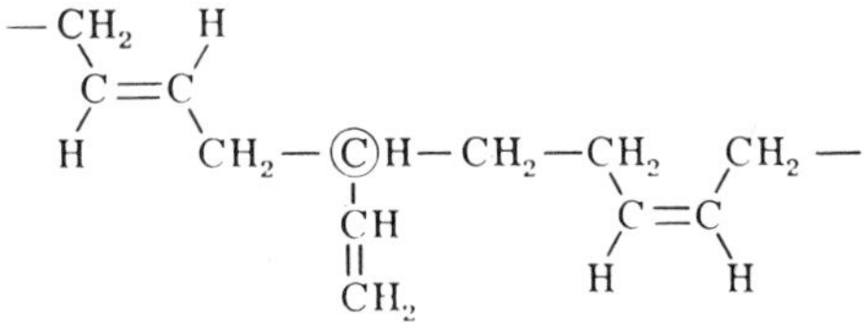 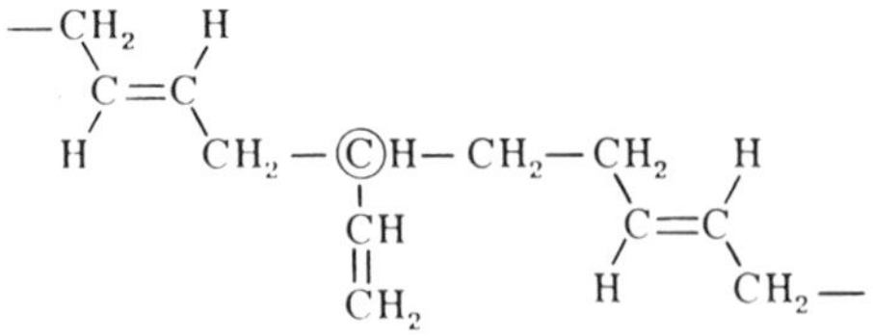

[a] See Fig. 4.5.
[b] Chemical shifts expressed in ppm referred to TMS for 20% (w/v) solution in $CDCl^3$ at 50°.

The overall composition of the polymer represented by Fig. 4.5 is 23% cis-1,4, 58% trans-1,4, 19% 1,2. Spectrum (c) is a computer simulation of (b) based on the assumption of a random distribution of units in these proportions. The fit is good, indicating that free radical propagation may be regarded as a Bernoullian process with regard to the generation of these isomeric structures, just as in the generation of pseudoasymmetric centers.

One of the most important uses of butadiene is in the manufacture of butadiene-based synthetic rubbers. These were developed in Germany prior to World War II. At first, experimentation centered about "Buna" rubbers made from butadiene using metallic sodium as catalyst, as the name implies. By 1930, I. G. Farben at Leverkusen had developed an aqueous emulsion system for making two new rubbers with superior properties: Buna-S, a copolymer with styrene, and Buna-N, a copolymer with acrylonitrile. Polybutadiene itself does not have sufficiently good mechanical properties, particularly tensile strength, to serve as a substitute for natural rubber. Styrene as comonomer improves its mechanical behavior; acrylonitrile also provides this improvement and confers oil resistance as well. Both copolymers contain about 75 wt% butadiene. In 1940, the U.S. Government became concerned over its rubber supply and made plans for larger scale production of Buna-S, renamed GR-S (for "Government rubber-styrene"), based primarily on the German technology. With the Japanese attack on Pearl Harbor and conquest of

Malaysia the synthetic rubber problem assumed crisis proportions. Full scale production of GR-S began in December, 1943, at Institute, West Virginia. Although total U.S. synthetic rubber production in 1940 was less than 2600 tons, it grew rapidly during the war, reaching 760,000 tons in 1946. By this time, many millions of tons had been produced. GR-S was a general purpose rubber, exceedingly important and useful although not equal to natural rubber for some uses, notably the treads of large tires. It is now known as SBR and plays a major role in the rubber market; natural rubber has not resumed its prewar dominance.

A significant American contribution to the technology of SBR, particularly in improving its tensile strength and mechanical properties generally, was the development of redox free radical initiating systems in which radical production can occur at practical rates in emulsion at 0° or even below (Bowden, 1979). Methanol is employed as an antifreeze. Most SBR rubber is now of this "cold" type, employing persulfate ion together with metals such as iron, chromium, and vanadium in low valence states. The principal effect of reduced temperature is to increase the regularity of the propagation; trans-1,4 enchainment predominates more strongly, as shown in Table 4.2

TABLE 4.2

The Effect of Temperature on the Structure of
Polybutadiene and Butadiene—Styrene Copolymer Chains

Polym. temp., °C	Styrene, wt.%	Composition of butadiene units		
		1,2,%	1,4,%	trans-1,4,%
−18	24.3	19.6	80.4	(83.9)
5	23.3	21.0	79.0	76.1
50	23.4	21.9	78.1	64.8
50	0	23.2	76.8	62.0
65	0	23.8	76.2	56.6
97	0	24.8	75.2	51.4

(composition determined by infrared spectroscopy). The composition of the butadiene sequences in the chains is not markedly affected by the presence of styrene.

With coordination catalysts, the structure of polybutadiene varies strongly with the composition of the catalyst, as shown in Table 4.3

TABLE 4.3

Catalysts for Stereospecific Polymerization of Butadiene[a]

Metal alkyl	Transition metal compound	Conditions	Structure of polymer (0/0)			
			1,2	cis-1,4	trans-1,4	
Et₂AlCl, EtAlCl₂	Cobalt compounds	Al/Co>100	—	95–98	—	
Et₃Al	VCl₄, VOCl₃	Al/V=2	—	—	95–100	
Et₃Al	TiI₄	Al/Ti=5	5	92	4	
Et₃Al	Chromium acetylacetonate	Al/Cr=10	Isotactic	—	—	
Et₃Al	Chromium acetylacetonate	Al/Cr=3	Syndiotactic	—	—	
EtAlCl₂	β-TiCl₃	Al/Ti=1–5	—	—	100	
Et₃Al	Ti(OBu)₄	Al/Ti=3	90	10	—	
Et₃Al	TiCl₄	Al/Ti-1	2	49	49	

[a] From Boor (1967).

[from Bowden (1979) and Boor (1967)]. Using the β crystalline form of titanium trichloride, the product is entirely trans-1,4. With a small proportion of cobalt compounds, less than 1 mol%, the structure is almost entirely cis-1,4. With chromium acetylacetonate the chain is 1,2 and can be switched from isotactic to syndiotactic by varying the aluminum/chromium ratio.

Rhodium salts are a particularly interesting class of coordination catalysts. These produce very highly stereospecific trans-1,4-polybutadiene (Rinehart, 1961) and can be employed in aqueous solution with a surfactant but without the need for an alkylating co-catalyst.

Anionic polymerization of butadiene with initiators such as alkali metal adducts of naphthalene and other polycyclic aromatic hydrocarbons yields a mixture of structures with trans-1,4 predominating. Lithium metal initiation produces a polybutadiene having *ca.* 90% 1,4 enchainment, 40% of which is cis. Other alkali metals produce chains with greatly increased vinyl content. Alkyl lithium initiators give products that depend upon the solvent employed. For example, with n — butyllithium (Duck and Locke, 1968), the addition of ether solvents such as diethyl ether, tetrahydrofuran, and (most effective) diethyleneglycol dimethyl ether ("diglyme") causes a sharp increase in 1,2 content from *ca.* 8% to as high as 85%. [A convenient review of these matters will be found in Richards (1977).]

4.3 POLYISOPRENE

It has been observed (Chen, 1962a,b; Golub *et al.*, 1962) that in carbon tetrachloride or carbon disulfide solution the methyl protons of trans-1,4-isoprene units are about 0.07 ppm more shielded than those of cis-1,4 units. In benzene the difference is slightly greater. In Fig. 4.6 (Bovey, 1972), the cis methyl appears at 1.63 ppm and the trans at 1.53 ppm. The methylene ($\sim$2.0 ppm) and vinylene protons ($\sim$5.2 ppm) are less sensitive to geometrical isomerism. Spectrum (c) in Fig. 4.6 is that of a polyisoprene containing about 80% of 3,4 units and 20% of 1,2 units (see Section 1.4). The 3,4-vinyl protons appear at 4.75 ppm. The 1,2-vinyl protons exhibit a peak at 4.95 ppm; intensity considerations indicate that a part of their resonance lies under the 3,4-vinyl proton peak. The most useful resonances for analytical purposes are the methyl peaks, appearing at 1.59 ppm for the pendant isopropenyl group of the 3,4 units and at 1.05 ppm for the 1,2 units, in which the methyl groups are not adjacent to double bonds.

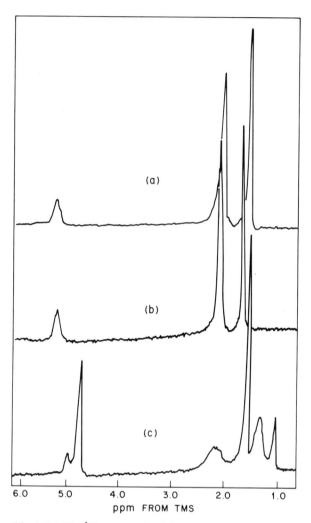

Fig. 4.6. The 100 MHz ^{1}H spectra of polyisoprenes using 5% solutions in C_6D_6 at 74°: (a) trans-1,4-polyisoprene; (b) cis-1,4-polyisoprene; (c) polyisoprene containing about 80% of 3,4 units and 20% of 1,2 units (Bovey, 1972).

In Fig. 4.7 are shown the 50 MHz spectra of cis- and trans-1,4-polyisoprene. They are similar to 15 MHz spectra originally reported by Duch and Grant (1970). There are five resonances, one for each isoprene carbon, assigned as indicated on the basis of model compounds. As with polybutadiene, the vinylene carbons are relatively insensitive to geometrical isomerism, but C_1 and the methyl

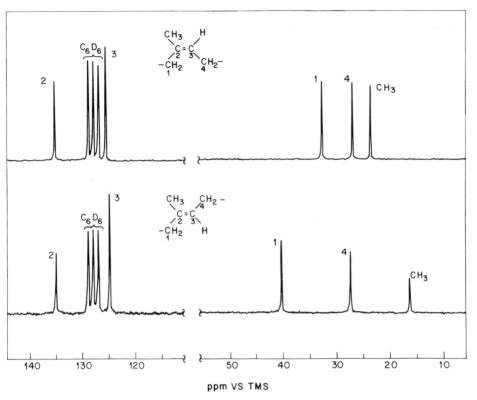

Fig. 4.7. The 50.3 MHz ^{13}C NMR spectra of (a) cis- and (b) trans-1,4-polyisoprene, observed in 10% (w/v) solution in C_6D_6 at 60°. (F. A. Bovey and F. C. Schilling, unpublished observation, 1981).

carbon are highly sensitive, showing again increased shielding when "seeing" a methylene or methyl carbon in a cis arrangement across the double bond. The C_4 carbon, which always sees another cis carbon, is insensitive to double bond geometry.

Spectrum (a) is that of natural rubber or *hevea brasiliensis*. The biochemical pathway to natural rubber is a known enzymatic process and is closely related to those by which terpenes and steroids are formed. Isoprene as such plays no part. The polymer is highly stereoregular, no trace of the trans structure being observable. Synthetic cis-1,4-polyisoprene has been manufactured on a substantial scale for many years by two processes, one employing lithium alkyls and the other a trialkylaluminum/titanium tetrachloride coordination catalyst. Both contain sufficient trans units (2—6%) to be readily observable by ^{13}C NMR, and are not regarded as fully equal to natural rubber for certain exacting processes such as the manufacture of truck and airplane tires. In most respects, they are essentially equivalent.

Spectrum (b) corresponds to *balata*, a naturally occurring trans-polyisoprene which is also highly stereoregular. Synthetic trans-polyisoprenes can be prepared with coordination catalysts based on vanadium, for example, triisobutylaluminum and VCl_3. Such products are at least 98% trans but do not command any market in the United States because of high price. Balata is not a rubber but rather a crystalline plastic; it too finds only a very limited market.

4.4 POLYCHLOROPRENE

Poly-(2-chlorobutadiene) or polychloroprene, more commonly known as Neoprene, was the earliest commercially successful synthetic rubber, being first made available by the duPont company in 1932. It has excellent mechanical properties and oil resistance. The preparation and polymerization of chloroprene were first described by Carothers and co-workers (1931, 1932). The monomer is prepared by the addition of hydrogen chloride to monovinylacetylene:

$$CH_2{=}CHC{\equiv}CH \ + \ HCl \xrightarrow[\substack{CuCl \\ NH_4Cl}]{30^\circ} CH_2{=}CHCCl{=}CH_2$$

The monovinylacetylene is prepared by the dimerization of acetylene, essentially as first described by Father Nieuwland of Notre Dame (Nieuwland *et al.*, 1931). Polymerization is carried out in emulsion with free radical initiators, for example, potassium persulfate. Several varieties of Neoprene are offered, some of which incorporate small amounts of comonomers. These will not concern us here.

Ozonolysis and infrared measurements (Walker and Mochel, 1948) indicated that ~95% of the units are trans-1,4. Later infrared studies, together with 60 MHz proton observations, led to the conclusion that the chains are more irregular than previously thought, the irregularities consisting mainly of head-to-head:tail-to-tail inversions (Ferguson, 1964). All-cis 1,4-polychloroprene can be prepared by chlorination of 1,4-poly-2-(tri-*n*-butyl)-1,3-butadiene, but it is found (Aufdermarsh and Pariser, 1964) that here too 20−25% of the monomer units are inverted.

As may be expected, the most detailed and revealing view of the chain structure of polychloroprene is obtained by ^{13}C NMR (F. C.

Schilling and F. A. Bovey, unpublished observations, 1976; Coleman *et al.*, 1977). As we have seen in Chapter 1, the following units may occur in the chains of an unsymmetrically substituted diene:

$$\begin{array}{cccc} cis\text{-}1,4 & trans\text{-}1,4 & 1,2 & 3,4 \end{array}$$

In addition, the occurrence of an isomerized 1,2 unit has been recognized:

$$-CH_2-\underset{\substack{\| \\ CH \\ | \\ CH_2Cl}}{C}-$$

Isomerized 1,2

In Fig. 4.8 is shown the 22.6 MHz ^{13}C spectrum of a Neoprene prepared in emulsion at an unspecified temperature. It is found by Coleman *et al.* (1977) that irregularities of all types increase with polymerization temperature. The principal resonances are labeled a, b, c, etc., in order of increasing shielding. The assignments are given in Table 4.4 and are taken from those of Coleman *et al.*, which were based on model compounds and statistical reasoning. The spectrum is divided into olefinic and aliphatic carbon regions, the first of which shows somewhat better resolved fine structure. In the olefinic region, triads of monomer units can be distinguished. Assuming that inversions of trans-1,4 units occur at random, the sequences

$$t\text{-}4,1\text{:}t\text{-}1,4\text{:}t\text{-}4,1$$

$$t\text{-}4,1\text{:}t\text{-}1,4\text{:}t\text{-}1,4$$

$$t\text{-}1,4\text{:}t\text{-}1,4\text{:}t\text{-}4,1$$

should occur with equal probability. The second and third are not resolved from each other and appear as a resonance twice as large as that of the first. Inverted cis-1,4 units probably also occur, but with a frequency too small to detect since the total content of cis units is only 4.8%.

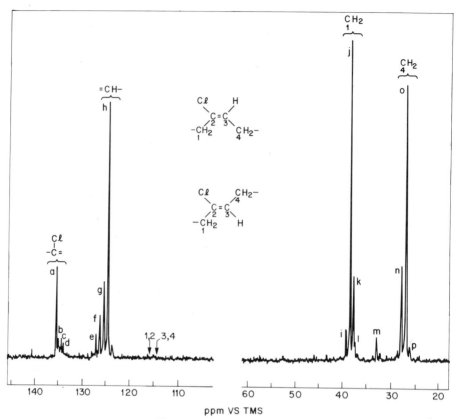

Fig. 4.8.   The 22.6 MHz ^{13}C NMR spectrum of Neoprene observed in 11.5% (w/v) solution in CDCl$_3$ at 48°C. (F. C. Schilling and F. A. Bovey, unpublished observations, 1976).

In the aliphatic region, only monomer dyads can be resolved. As might be expected, C$_1$ carbons are sensitive to the geometry of the unit to the left and C$_4$ carbons are sensitive to the unit to the right, as conventionally represented.

About 12.8% of the trans-1,4 units are inverted. Assuming random statistics, approximately 1.6% of pairs of inverted units should occur, but it is not clear that these would be detectable. The occurrence of 1,2 and 3,4 units is so infrequent that their resonances cannot be observed on the scale of Fig. 4.8. Measurement shows about 1.9% of 1,2 units and 1.0% of 3,4 units.

Cais and Stuk (1980) have carried out a very thorough analysis by ^{13}C NMR of the structure of chloroprene—sulfur dioxide copolymers.

Geometrical isomerism in these copolymers is very similar to that in the homopolymer and showed an approximately equal temperature dependence.

TABLE 4.4

^{13}C Peak Assignments for Neoprene[a]

Peak designation	Chemical shift	Assignment[a]
a	135.1	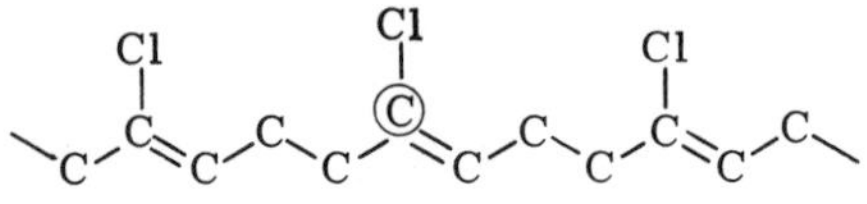 (structure) t-1,4 t-1,4 t-1,4
b,c,d		Other triads centered on 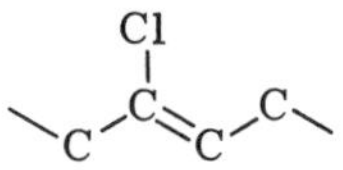 (structure)
e	126.7	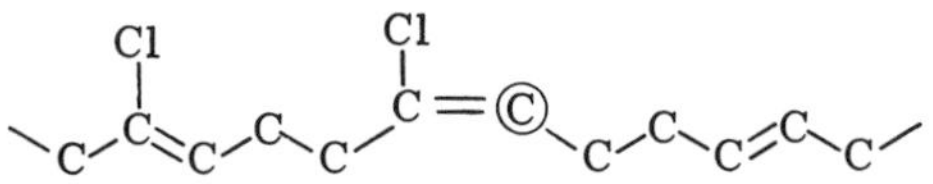 (structure) t-1,4 c-1,4 t-1,4
f	125.9	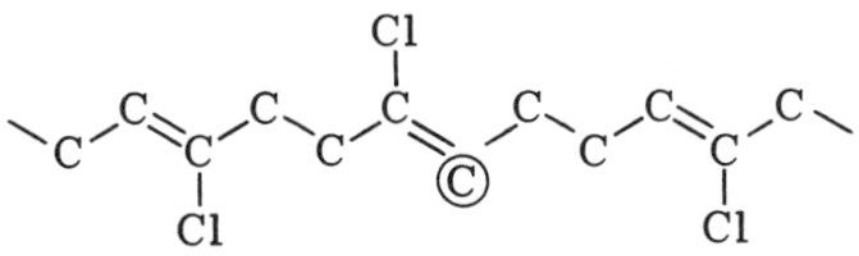 (structure) t-4,1 t-1,4 t-4,1
g	125.0	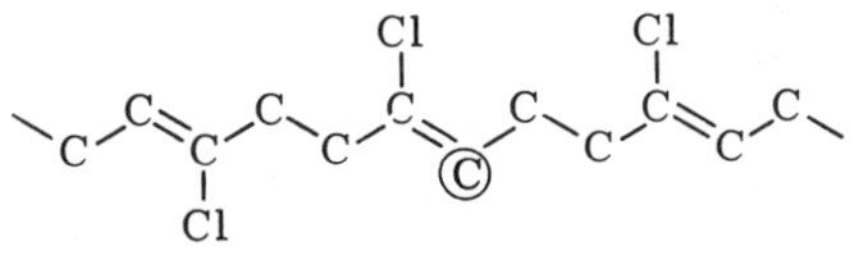 (structure) t-4,1 t-1,4 t-1,4

TABLE 4.4, CONTINUED

Peak designation	Chemical shift	Assignment[a]
g, continued		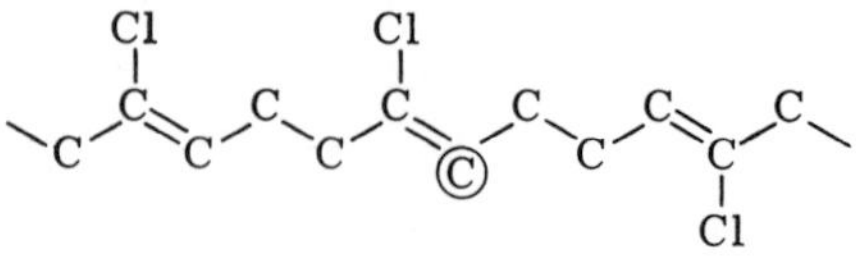 t-1,4 t-1,4 t-4,1
h	124.2	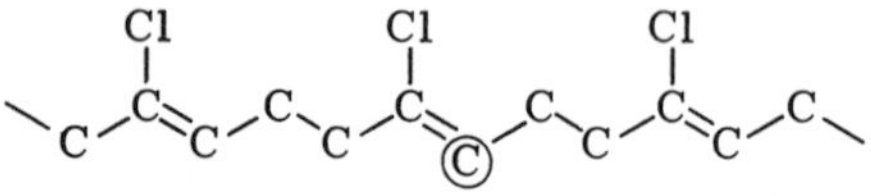 t-1,4 t-1,4 t-1,4
i	39.2	Unassigned
j	38.4	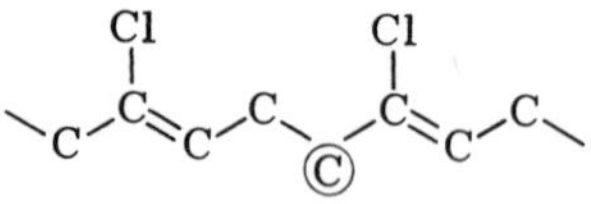 t-1,4 t-1,4
k	37.7	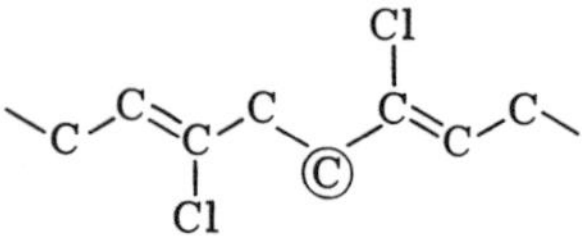 t-4,1 t-1,4
l	37.5	Unassigned
m	32.8	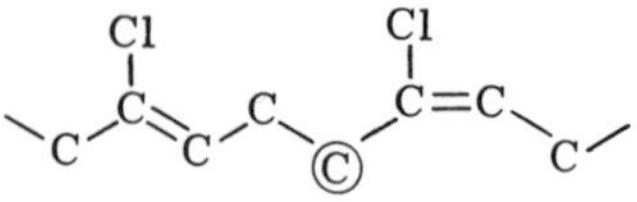 t-1,4 c-1,4

TABLE 4.4, CONTINUED

Peak designation	Chemical shift	Assignment[a]

n   27.7

$$\begin{array}{c} Cl \\ | \\ \diagdown C\diagdown C\!\!=\!\!C\diagdown ©\diagdown C\diagdown C\!\!=\!\!C\diagdown C\diagdown \\ | \\ Cl \end{array}$$

t-1,4   t-4,1

o   26.9

$$\begin{array}{c} Cl \qquad\quad Cl \\ | \qquad\qquad | \\ \diagdown C\diagdown C\!\!=\!\!C\diagdown ©\diagdown C\diagdown C\!\!=\!\!C\diagdown C\diagdown \end{array}$$

t-1,4   t-1,4

p   26.0   Unassigned

[a] t: *trans*; c: *cis*.

REFERENCES

Aufdermarsh, C. A., and Pariser, R. (1964). *J. Polymer Sci., Part A* **2**, 4727.
Boor, J., Jr. (1967). *Macromol. Rev.* **2**, 115.
Bovey, F. A. (1972). *In* "High Resolution NMR of Macromolecules," p. 220. Academic Press, New York.
Bowden, M. J. (1979). *In* "Macromolecules" F. A. Bovey, and F. H. Winslow, eds., pp. 30-32. Academic Press, New York.
Cais, R. E., and Stuk, G. J. (1980). *Macromolecules* **13**, 415.
Carothers, W. H., Williams, I., Collins, A. M., and Kirby, J. E. (1931). *J. Am. Chem. Soc.* **53**, 4203.
Carothers, W. H., Berchet, G. J., and Collins, A. M. (1932). *J. Am. Chem. Soc.* **54**, 4066.
Chen, H. Y. (1962a). *Anal. Chem.* **34**, 1134.
Chen, H. Y. (1962b). *Anal. Chem.* **34**, 1793.
Clague, A. D. H., Vanbroek, J. A., and Blaauw, L. P. (1974). *Macromolecules* **7**, 348.
Coleman, M. M., Tabb, D. L., and Brame, E. G. (1977). *Rubber Chem. Technol.* **50**, 49.
Conti, F., Segré, A.-L., Pini, P., and Porri, L. (1974a). *Polymer* **15**, 5.
Conti, F., Delfini, M., Segré, A.-L., Pini, D., and Porri, L. (1974b). *Polymer* **15**, 816.
Duch, M. W., and Grant, D. M. (1970). *Macromolecules* **3**, 165.
Duck, E. W., and Locke, M. (1968). *J. Inst. Rubber Ind. (A)* **6**, 3407.
Elgert, K. F., Quack, G., and Stützel B. (1974). *Makromol. Chem.* **15**, 612.
Ferguson, R. C. (1964). *J. Polymer Sci., Part A* **2**, 4735.
Furukawa, J., Kobayashi, E., Katsuki, N., and Kawagoe, T. (1974). *Makromol. Chem.* **175**, 237.

Golub, M. A., Fuqua, S. A., and Bhacca, N. S. (1962). *J. Am. Chem. Soc.* **84**, 4981.

Grant, D. M., and Paul, E. G. (1964). *J. Am. Chem. Soc.* **86**, 2984.

Hatada, K., Tanaka, Y., Terawaki, Y., and Okuda, H. (1973). *Polymer J.* **5**, 327.

Julemont, M., Walckiers, E., Warin, R., and Teyssié, P. (1974). *Makromol. Chem.* **175**, 1673.

Mochel, V. D. (1972). *J. Polymer Sci.* **10**, 1009.

Nieuwland, J. A., Calcott, W. S., Downing, F. B., and Carter, A. S. (1931). *J. Am. Chem. Soc.* **53**, 4197.

Richards, D. H. (1977). *Chem. Soc. Rev.* **6**(2), 235-260.

Rinehart, R. E. (1961). *J. Am. Chem. Soc.* **83**, 4864.

Santee, E. R., Jr., Chang, R, and Morton, M. (1973a). *J. Polymer Sci. Polymer Lett. Ed.* **11**, 449.

Santee, E. R., Jr., Mochel, V. D., and Morton, M. (1973b). *J. Polymer Sci. Polymer Lett. Ed.* **11** (1973).

Tanaka, Y., Sato, H., Ogawa, M., Hatada, K., and Terawaki, Y. (1974). *J Polymer Sci., Polymer Lett. Ed.* **12**, 369.

Tanaka, Y., Sato, H., Hatada, K., Terawaki, Y., and Okuda, H. (1977). *Makromol. Chem.* **178**, 1823.

Walckiers, E., and Julemont, M. (1981). *Makromol. Chem.* **182**, 1541.

Walker, H. W., and Mochel, W. E. (1948). *Proc. 2nd Rubber Tech. Conf., London,* 1948, p. 69.

Chapter 5

COPOLYMERIZATION AND COPOLYMER STRUCTURE

5.1 INTRODUCTION

We have seen (Section 1.7) that in addition to the forms of isomerism we have discussed in Chapters 3 and 4, a very important structural variable is provided by our ability to synthesize not only homopolymers with a single type of monomer unit but also *copolymers* having chains composed of two or more comonomer units. (Indeed, homopolymers in which stereochemical configuration and geometrical isomerism can occur may be regarded as copolymers of differing units, although their composition is not so readily controlled as in true copolymerization.) Copolymers may be broadly divided into three types, which are shown in Fig. 1.2b: (a) random; (b) block; (c) graft. Block and graft copolymers contain relatively long sequences of one monomer bonded to similar sequences of another. They are of much scientific and technological interest, but we shall not discuss them here. Our attention will be confined to the "random" type, in which two or more types of comonomer units are present in each chain (although rarely with truly random sequence.) Although copolymers are of very broad occurrence and importance—proteins and nucleic acids are copolymers with specific sequences—we shall further confine our discussion to copolymers of vinyl and diene comonomers. (Exceptions are the segmented copolyesters described in Chapter 8.)

123

5.2 COPOLYMER COMPOSITION; THE COPOLYMER EQUATION

A very wide variety of monomers will copolymerize with one another, including some that do not polymerize alone. For example, sulfur dioxide copolymerizes readily with a variety of olefins by a free radical mechanism to form polysulfones with a rigorously alternating structure:

$$RCH{=}CH_2 \; + \; SO_2 \; \xrightarrow[\text{copolymerization}]{\text{free radical}} \; \left(\begin{array}{c} R \quad\;\; O \\ | \quad\quad \| \\ CHCH_2{-}S{-} \\ \| \\ O \end{array} \right)_x$$

Neither sulfur dioxide nor the olefins (with the exception of ethylene) homopolymerize by a free radical mechanism, yet the copolymerization occurs very rapidly.

Copolymerization can be employed to yield products with a range of physical and mechanical properties. Amorphous homopolymers consist of a single phase, but the phase structure of a copolymer depends on the distribution of comonomer units. Random copolymers usually form homogeneous phases with properties intermediate between those of the corresponding homopolymers. Block and graft copolymers, however, form more complex multiphase structures because in general the long chains of each type of monomer are not compatible with each other and so must exist in separate domains.

In the subsequent discussion we shall consider only the formation of homogeneous copolymers made (in principle, at least) by the addition of an appropriate initiator to a mixture of two monomers. The composition of a copolymer is not in general the same as that of the monomer mixture from which it is formed and cannot be deduced from the relative rates of homopolymerization of the monomers. It depends on the competing and differing reactivities of the monomers and the growing chain ends. In the treatment to follow, we shall designate the latter by a noncommittal asterisk, which may indicate a *free radical, anion, cation,* or *coordination complex* (as in Ziegler−Natta catalysis). Most of the work on copolymerization, especially in the "classical" era of exploration of this field in the 1940−1960 period, was actually done on free radical systems.

We need not consider in detail the *initiation* and *termination* steps in the polymerization reaction, as it is the *propagation* process that determines the chain composition and distribution of comonomer

units.   Four propagation steps must be considered in the
copolymerization of two monomers. These are shown below together
with the appropriate rate expression, formulated in terms of
consumption of monomer:

$$\text{\char`\~\char`\~\char`\~} M_1^* + M_1 \xrightarrow{\quad k_{11} \quad} \text{\char`\~\char`\~} M_1^*$$

$$-d[M_1]/dt = k_{11}[M_1^*][M_1]$$

$$\text{(5.1)}$$

$$\text{\char`\~\char`\~} M_1^* + M_2 \xrightarrow{\quad k_{12} \quad} \text{\char`\~\char`\~} M_2^*$$

$$-d[M_2]/dt = k_{12}[M_1^*][M_2]$$

$$\text{(5.2)}$$

$$\text{\char`\~\char`\~} M_2^* + M_1 \xrightarrow{\quad k_{21} \quad} \text{\char`\~\char`\~} M_1^*$$

$$-d[M_1]/dt = k_{21}[M_2^*][M_1]$$

$$\text{(5.3)}$$

$$\text{\char`\~\char`\~} M_2^* + M_2 \xrightarrow{\quad k_{22} \quad} \text{\char`\~\char`\~} M_2^*$$

$$-d[M_2]/dt = k_{22}[M_2^*][M_2]$$

$$\text{(5.4)}$$

The total rate of disappearance of M_1 is given by

$$-d[M_1]/dt = k_{11}[M_1^*][M_1] + k_{21}[M_2^*][M_1] \qquad \text{(5.5)}$$

and that of M_2 by

$$-d[M_2]/dt = k_{22}[M_2^*][M_2] + k_{12}[M_1^*][M_2] \qquad \text{(5.6)}$$

Dividing Eq. (5.5) by Eq. (5.6) gives the ratio of the rates at which the
two monomers enter the copolymer, which in turn must represent the
composition of the copolymer being formed at any instant:

$$\frac{d[M_1]}{d[M_2]} = \frac{[M_1]}{[M_2]}\left\{\frac{k_{11}[M_1^*]+k_{21}[M_2^*]}{k_{22}[M_2^*]+k_{12}[M_1^*]}\right\} \tag{5.7}$$

The ratio $[M_1]/[M_2]$ is, of course, a measure of the composition of the monomer mixture or *feed*. The terms in the reacting species M_1^* and M_2^*, the concentrations of which are very low and troublesome to establish, can be eliminated by invoking a *stationary state* condition for them. The existence of such a state implies that the rate of disappearance of M_1^* is equal to the rate of its conversion to M_2^* and vice versa:

$$k_{21}[M_2^*][M_1] = k_{12}[M_1^*][M_2] \tag{5.8a}$$

so that

$$[M_2^*] = k_{12}[M_2][M_1^*]/k_{21}[M_1] \tag{5.8b}$$

Substitution for $[M_2^*]$ in Eq. (5.7) gives

$$\frac{d[M_1]}{d[M_2]} = \frac{[M_1]}{[M_2]}\left\{\frac{k_{11}[M_1^*]+k_{12}[M_1^*]([M_2]/[M_1])}{(k_{22}k_{12}/k_{21})([M_2][M_1^*]/[M_1])+k_{12}[M_1^*]}\right\} \tag{5.9}$$

Dividing numerator and denominator by $k_{12}[M_1^*]$ we have

$$\frac{d[M_1]}{d[M_2]} = \frac{[M_1]}{[M_2]}\left\{\frac{(k_{11}/k_{12})+([M_2]/[M_1])}{(k_{22}/k_{21})([M_2]/[M_1])+1}\right\} \tag{5.10}$$

$$= \frac{[M_1]}{[M_2]}\left\{\frac{(k_{11}/k_{12})[M_1]+[M_2]}{(k_{22}/k_{21})[M_2]+[M_1]}\right\} \tag{5.11}$$

[It has been shown (Melville *et al.*, 1947; Goldfinger and Kane, 1948) that this same result can be obtained by statistical arguments without the need to assume a steady state.]

We now define *reactivity ratios* as

$$r_1 = k_{11}/k_{12} \tag{5.12}$$

the rate of reaction of the active center M_1^* with monomer 1 compared to its reaction with monomer 2; and correspondingly

$$r_2 = k_{22}/k_{21} \qquad (5.13)$$

the ratio of the rate of reaction of M_2^* with monomer 2 to its rate with monomer 1. We can then rewrite Eq. (5.11) as

$$\frac{d[M_1]}{d[M_2]} = \frac{[M_1]}{[M_2]}\left\{\frac{r_1[M_1]+[M_2]}{r_2[M_2]+[M_1]}\right\} \qquad (5.14)$$

This relationship between instantaneous copolymer composition and monomer feed composition is known as the *copolymer equation*. The main purpose of a study of any copolymer system is to obtain data that may be interpreted in terms of this equation to provide the reactivity ratios r_1 and r_2. If these are established, then for that system the instantaneous copolymer composition can be calculated for any monomer feed.

It is usually more convenient to express monomer concentration as *mole fraction* in both the feed and the copolymer. The feed mole fraction is given (for monomer 1) by

$$f_1 = 1 - f_2 = \frac{[M_1]}{[M_1]+[M_2]} \qquad (5.15)$$

The instantaneous copolymer composition is given by

$$F_1 = 1 - F_2 = \frac{d[M_1]}{d[M_1]+d[M_2]}$$

Equation (5.14) can then be recast as

$$F_1 = \frac{r_1 f_1^2 + f_1 f_2}{r_1 f_1^2 + 2f_1 f_2 + r_2 f_2^2} \qquad (5.16)$$

This is the most generally useful form of the copolymer equation.

It must be emphasized again that the copolymerization equation describes the instantaneous composition of the copolymer. Since the comonomers generally do not enter the polymer in the same proportion as in the feed, the latter will drift in composition, becoming depleted in the more reactive monomer. As a result, the higher the monomer conversion the more heterogeneous the product. It is therefore customary in fundamental studies of copolymer systems to limit the conversion to about 5% or less, although drifts in composition can be dealt with mathematically (see Bowden, 1979). In copolymer production on a practical scale, it is customary to achieve greater regularity by adjusting the monomer input; this usually means withholding the more reactive monomer.

5.3 REACTIVITY RATIOS

The significance of reactivity ratios can best be appreciated by considering first some special cases.

Case i. $r_1 = 1/r_2$ or $r_1 r_2 = 1$.

Let us suppose that the ratio of the rate of reaction of M_1^* with M_1 compared to that of M_1^* with M_2, represented by k_{11}/k_{12}, is the same as the ratio of the rate of reaction of M_2^* with M_1 compared to that of M_2^* with M_2, given by k_{21}/k_{22}. That is

$$k_{11}/k_{12} = k_{21}/k_{22}$$

or

$$r_1 = 1/r_2,$$

i.e.,

$$r_1 r_2 = 1$$

If this holds true it means that the nature of the chain end, whether M_1^* or M_2^*, has *no influence on the relative rates of addition of the monomers*. Such a copolymerization system is termed *ideal* or *random*. Plots of a number of cases, with varied values of r_1, are shown in Fig. 5.1.

A special case is given by $r_1 = r_2 = 1$. This "super ideal" case is represented as the straight line of 45° slope in Fig. 5.1, and corresponds to a system in which both monomers show equal reactivity toward both M_1^* and M_2^*. No such actual case is known, but one or two systems approximate it. (Isotopically labeled monomers would usually come very close.) In such a system, the polymer composition is always equal to the monomer composition, and this remains true throughout the reaction.

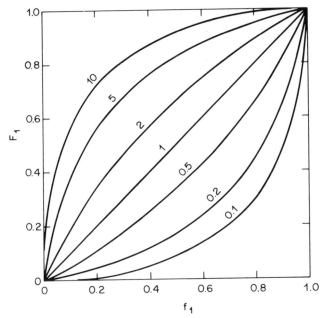

Fig. 5.1. Instantaneous composition of copolymer (mole fraction F_1) as a function of monomer composition (mole fraction f_1) for the values of r_1 indicated on the curves. In all cases, $r_1 r_2 = 1$.

Case ii. $r_1 = r_2 = 0$.

Under these conditions, neither reactive species can react with its own monomer but only with the other, i.e., M_1^* can react only with M_2 and M_2^* only with M_1. The copolymerization equation reduces to

$$d[M_1]/d[M_2] = 1 \qquad (5.17)$$

or

$$F_1 = F_2 = 0.5 \qquad (5.18)$$

regardless of the value of f_1. The plot of F_1 as a function of f_1 is a horizontal line at $F_1 = 0.5$, as shown in Fig. 5.2d. It follows that the monomer units must enter the chain in strict alternation:

$$\cdots M_1 M_2 M_1 M_2 M_1 M_2 M_1 M_2 \cdots$$

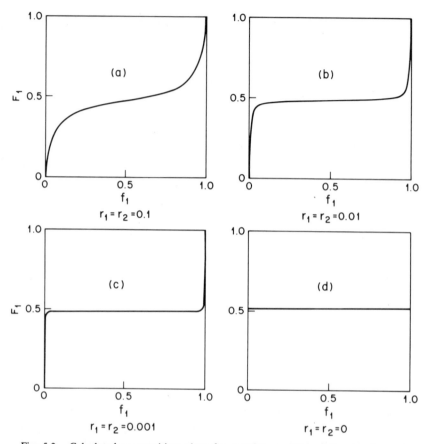

Fig. 5.2. Calculated composition plots for copolymer systems showing increasing tendency toward alternation.

This seemingly exaggerated situation is actually of fairly common occurrence. We have already seen an example in the copolymerization of olefins with sulfur dioxide (p. 124).

Many systems show a tendency toward alternating behavior in which r_1 and r_2 are both markedly less than 1. A few cases are illustrated in Fig. 5.2. We shall discuss some individual examples of this behavior in Section 5.5.

Case iii. $r_1, r_2 > 1$.

Each reacting chain end type prefers to add its own monomer, leading to the formation of *block sequences*:

$$\cdots M_1 M_1 M_1 M_1 M_2 M_2 M_2 M_1 M_1 M_1 \cdots$$

It is common for one monomer of a pair to prefer its own growing species, as we have seen (Fig. 5.1), but for reasons not entirely clear the case contemplated here, where both show such a preference, is very rare. Figure 5.3 shows calculated plots for hypothetical systems in which $r_1 = r_2$ and both substantially exceed one. An odd circumstance about these plots does not seem to have been pointed out in previous discussions and may be worth mentioning. Note that when r_1 and r_2 exceed 10, the curves become indistinguishable, and do not have as a limit a vertical straight line at $f_1 = 0.5$, by analogy to the limiting case when $r_1 = r_2 = 0$ (Fig. 5.2d). It can be seen from Eq. (5.16) why this is so. When r_1 and r_2 are equal and very large, then $f_1 f_2$ can be neglected in comparison with the other terms, and we find that

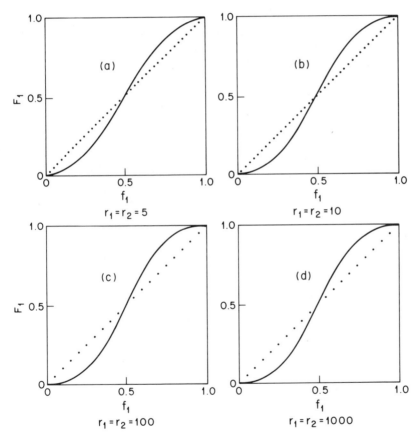

Fig. 5.3. Calculated composition plots for copolymerization systems with $r_1 = r_2 > 1$, i.e., with a "blocklike" tendency.

$$F_1 = \frac{f_1^2}{2f_1^2 - 2f_1 + 1} \qquad (5.19)$$

that is, F_1 is independent of the reactivity ratios. The chemical significance of this result, if any, is not clear. A case of simultaneous independent homopolymerizations appears to be that of oxetane and N-vinylcarbazole, initiated by tetranitromethane (Gumbs *et al.*, 1969).

The most common behavior of monomer pairs is copolymerization characterized by products of reactivity ratios between the extremes just discussed, i.e.

$$0 < r_1 r_2 < 1$$

A set of nonideal composition plots with r_1 fixed at 0.5 and r_2 varying from 0 to 4.0 is shown in Fig. 5.4. These curves show the effect of an increasing tendency toward alternation as the product $r_1 r_2$ approaches zero. This point has already been illustrated by the curves in Fig. 5.2, where r_1 and r_2 are equal and less than 1. When r_1 and r_2 are both

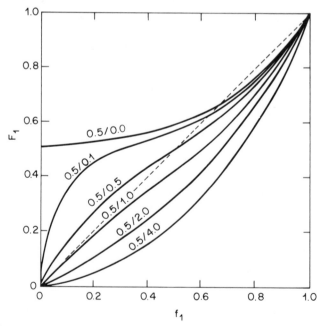

Fig. 5.4.   Calculated composition plots for copolymerization systems for $r_1 = 0.5$ and varied r_2 as shown. Numbers on curves are r_1/r_2.

less than 1, whether equal or not, the composition curve crosses the diagonal representing $F_1 = f_1$. At the crossing, the composition of feed and copolymer are equal, and at this point *azeotropic copolymerization* is said to occur, by analogy to boiling liquid mixtures exhibiting parallel behavior. When a system attains this point its composition will not change further. (For a typical alternating system, if f_1 is less than the azeotropic composition, the system will move away from this point; if f_1 exceeds the azeotropic composition, the system will move toward it.)

Solution of the copolymer equation for the azeotropic condition, i.e.

$$\frac{d[M_1]}{d[M_2]} = \frac{[M_1]}{[M_2]} \tag{5.20}$$

gives

$$\frac{[M_1]}{[M_2]} = \frac{(1-r_2)}{(1-r_1)} \tag{5.21}$$

or for the critical feed composition

$$(f_1)_c = \frac{(1-r_2)}{(2-r_1-r_2)} \tag{5.22}$$

5.4 DETERMINATION OF REACTIVITY RATIOS

At the time when copolymerization first became a topic of major interest—in the nineteen-forties and fifties—there was no effective means at hand for the determination of the microstructure of copolymers. The only recourse was the measurement of the monomer ratio (usually by elemental analysis) in copolymers prepared from varied feed ratios. These data were then plotted in various ways to extract r_1 and r_2. The most straightforward procedure is simply to plot F_1 as a function of f_1 in the way already indicated. The values of r_1 and r_2 that best fit this plot are then established by trial. [Approximate values may be estimated from the slope of the plot near $f_1 = 0$. When $f_1 \to 0$ it of course follows that $f_2 \to 1$ and therefore, from Eq. (5.16),

$$F_1 \rightarrow f_1/r_2$$

or

$$F_1/f_1 \rightarrow 1/r_2,$$

i.e, $1/r_2$ is approximated by the tangent of the angle of the composition curve at the origin. Similarly, the upper right corner of the plot provides an estimate of $1/r_1$.] The direct curve-fitting procedure is tedious and can only give approximate values of the reactivity ratios, since the composition curve is rather insensitive to small changes in r_1 and r_2.

The copolymer Eq. (5.14) can be rearranged to solve for one of the reactivity ratios (Mayo and Lewis, 1944):

$$r_2 = \frac{[M_1]}{[M_2]} \left\{ \frac{d[M_2]}{d[M_1]} \left(1 + r_1 \frac{[M_1]}{[M_2]} \right) - 1 \right\} \qquad (5.23)$$

and a parallel expression for r_1. From a particular pair of feed and copolymer compositions, one can obtain values of r_2 for assumed values of r_1. A plot of r_1 and r_2 according to Eq. (5.23) gives a straight line. The same procedure is used for other pairs of compositions, resulting in a family of straight lines which in the absence of experimental error should intersect at a point in the $r_1 r_2$ plane. In practice the lines actually intersect over an area, and various procedures have been proposed for selecting the best point within this area.

Another procedure is that of Fineman and Ross (1950). For this purpose, Eq. (5.16) is rearranged to

$$\frac{(1/F_2-2)}{(1/f_2-1)} = r_1 - r_2 \frac{(1/F_2-1)}{(1/f_2-1)^2} \qquad (5.24)$$

Such a plot of the data for the system N-vinylsuccinimide(1)-methyl acrylate(2) is shown in Fig. 5.5. One may also plot the data according to

$$\frac{(1/f_2-1)(1/F_2-2)}{(1/F_2-1)} = -r_2 + r_1 \frac{(1/f_2-1)^2}{(1/F_2-1)} \qquad (5.25)$$

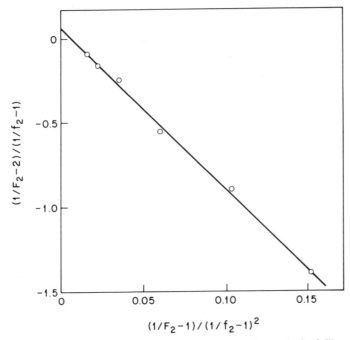

Fig. 5.5.   Reactivity ratio determination according to the method of Fineman and Ross for the system N-vinylsuccinimide(1)-methyl acrylate(2)-; $r_1 = 0.07$ and $r_2 = 9.7$. [Data from H. Hopff and P. C. Schlumbom, cited in Elias (1977).] In this plot, r_1 is given by the intercept on the ordinate.

Such a plot of the same data is shown in Fig. 5.6. In the first plot, r_1 is obtained from the intercept on the ordinate (more accurately estimated than the slope), while in the second plot r_2 is so obtained. The values are

$$r_1 = 0.07$$

$$r_2 = 9.3$$

or

$$r_1 r_2 = 0.65$$

These results show that methyl acrylate is much more reactive toward its own radicals than is N-vinylsuccinimide and is also much more reactive than N-vinylsuccinimide toward N-vinylsuccinimide free radicals. The system is similar to that represented in Fig. 5.1 for $r_1 = 0.1$, and approximates to an "ideal" case (p. 128).

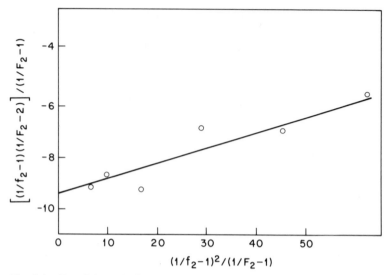

Fig. 5.6.   Reactivity ratio determination for the system shown in Fig. 5.5. Here r_2 is given by the intercept on the ordinate.

TABLE 5.1

Reactivity Ratios for Copolymerization at 60°C

M_1	M_2	r_1	r_2
Styrene	Acrylonitrile	0.4 ± 0.05	0.04 ± 0.04
Styrene	Methyl methacrylate	0.52 ± 0.02	0.46 ± 0.02
Styrene	Butadiene	0.78 ± 0.01	1.39 ± 0.03
Styrene	Vinyl acetate	55 ± 10	0.01 ± 0.01
Styrene	Maleic anhydride	0.02	0
Methyl methacrylate	Acrylonitrile	1.2 ± 0.14	0.15 ± 0.07
Methyl methacrylate	Vinyl acetate	20 ± 3	0.015 ± 0.015
Methyl methacrylate	Methyl acrylate	1.69	0.34
Vinyl acetate	Acrylonitrile	0.061 ± 0.013	4.05 ± 0.3
Vinyl acetate	Vinyl chloride	0.23 ± 0.02	1.68 ± 0.08
Vinylidene chloride	Isobutene	3.3	0.05

In Table 5.1, reactivity ratios for free radical copolymerization of a number of carefully studied monomer pairs are listed in order to furnish an indication of the range of behavior observed. These reactivities have been determined by traditional methods but in some cases confirmed by sequence analysis (next section).

5.5 MONOMER SEQUENCES AND THEIR OBSERVATION

The computational and graphical methods devised to allow the determination of reactivity ratios with optimum accuracy from overall composition data are tedious and not entirely satisfactory. This is basically a rather crude and insensitive approach to the problem. It would be very advantageous to be able to observe and measure the *comonomer sequences* directly. The theoretical treatment of the copolymerization process that predicts the overall composition also predicts the frequency of occurrence of specific comonomer sequences. This was realized very early but, as we have said, there was at that time no way to observe the sequences. This can now be readily done by ^1H and ^{13}C NMR, which in addition allows one to observe the *copolymer stereochemistry* (never considered in earlier work) and the presence of anomalous units (Chapter 6). It also permits one to detect deviations from the simple "terminal" model used in the discussion so far, in which one assumes that only the end monomer unit of the growing chain determines its reactivity. Such deviations can be detected also by the classical approach, but this is much more cumbersome. Finally, it should be noted that by sequence measurements one can determine reactivity ratios from only a single copolymer, although of course the feed ratio must be known. It may still be desirable to observe a range of compositions to assist in resonance assignments but it is not in principle essential.

The NMR analysis of copolymer structure, like that of stereochemical configuration, can be carried to varying depths of complexity, depending on the lengths of the sequences that can be observed. If the copolymerization is not strictly alternating or blocklike, the dyad, triad, and tetrad sequences may be represented as follows, stereochemistry being ignored:

Dyads: m_1m_1 $m_1m_2(\text{or } m_2m_1)$ m_2m_2

Triads: $m_1m_1m_1$ $m_2m_2m_2$

 $m_1m_1m_2(\text{or } m_2m_1m_1)$ $m_1m_2m_2(\text{or } m_2m_2m_1)$

 $m_2m_1m_2$ $m_1m_2m_1$

Tetrads: $m_1m_1m_1m_1$ $m_1m_1m_2m_1(m_1m_2m_1m_1)$ $m_2m_2m_2m_2$

$m_1m_1m_1m_2(m_2m_1m_1m_1)$ $m_1m_1m_2m_2(m_2m_2m_1m_1)$ $m_2m_2m_2m_1(m_1m_2m_2m_2)$

$m_2m_1m_2m_1(m_1m_2m_1m_2)$

$m_2m_1m_1m_2$ $m_2m_1m_2m_2(m_2m_2m_1m_2)$ $m_1m_2m_2m_1$

The dyad probabilities are given by

$$[m_1m_1] = F_1P_{11} \tag{5.26}$$

$$[m_1m_2] \text{ (or } [m_2m_1]) = 2F_1(1-P_{11}) \tag{5.27}$$

since

$$P_{12} = 1 - P_{11} \tag{5.28}$$

etc.,

$$[m_2m_2] = F_2P_{22} \tag{5.29}$$

Here, as in the previous discussion, F_1 and F_2 are the overall mole fractions of monomers 1 and 2 in the polymer, i.e., the *unconditional probability* of finding each monomer in a very long polymer chain. These can be determined from the NMR spectrum. The quantity P_{11} expresses the *conditional probability* that a chain ending in m_1 will add another m_1 and P_{22} expresses the corresponding probability for m_2. P_{12} is the probability that a chain ending in m_1 will add m_2, equal to the probability that it will *not* add m_1, or $1 - P_{11}$. There are thus four probabilities, P_{11}, P_{12}, P_{21}, and P_{22}, but, as in the analogous treatment of stereochemical probabilities (p. 52), they are related by

$$P_{11} + P_{12} = 1$$

$$P_{21} + P_{22} = 1$$

We choose to employ P_{11} and P_{22}; it can be shown that they are given by

$$P_{11} = \frac{r_1f_1}{1-f_1(1-r_1)} \tag{5.30}$$

$$P_{22} = \frac{r_2 f_2}{1 - f_2(1 - r_2)} \qquad (5.31)$$

From these relationships we have

$$r_1 = \frac{(1 - f_1)[m_1 m_1]}{f_1(F_1 - [m_1 m_1])} \qquad (5.32)$$

and

$$r_2 = \frac{(1 - f_2)[m_2 m_2]}{f_2(F_2 - [m_2 m_2])} \qquad (5.33)$$

It is evident from these expressions that, as already stated, r_1 and r_2 may be obtained from a single polymer provided the feed composition is known. A completely anonymous polymer without any pedigree can still be effectively analyzed but cannot yield values of the reactivity ratios.

Entirely analogous expressions apply to triad and tetrad sequences. For example, for m_1-centered triads

$$[m_1 m_1 m_1] = F_1 P_{11}^2 \qquad (5.34)$$

$$[m_1 m_1 m_2] \text{ (or } [m_2 m_1 m_1]) = 2F_1 P_{11}(1 - P_{11}) \qquad (5.35)$$

$$[m_2 m_1 m_2] = F_2(1 - P_{22})(1 - P_{11}) \qquad (5.36)$$

and for $m_1 m_1$-centered tetrads

$$[m_1 m_1 m_1 m_1] = F_1 P_1^3 \qquad (5.37)$$

$$[m_1 m_1 m_1 m_2] \text{ (or } [m_2 m_1 m_1 m_1]) = 2F_1 P_{11}^2 P_{12} \qquad (5.38a)$$

$$= 2F_1 P_{11}^2 (1 - P_{11}) \qquad (5.38b)$$

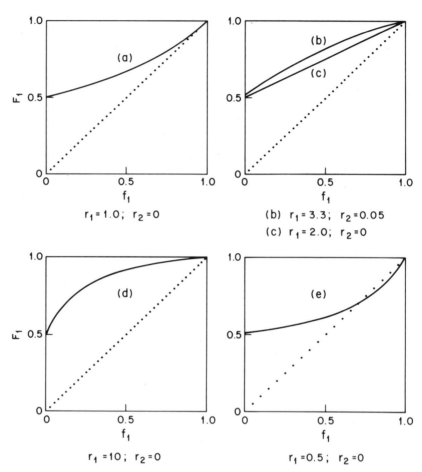

Fig. 5.7. Calculated composition plots for copolymerization systems with varied values of r_1 and r_2 equal to zero. Plot (b) represents the system vinylidene chloride(1): isobutylene(2), for which $r_1 = 3.3$ and $r_2 = 0.05$.

$$[m_2m_1m_1m_2] = F_2P_{11}(1-P_{11})(1-P_{22}) \qquad (5.39)$$

We consider first the free radical copolymerization of vinylidene chloride (M_1) and isobutylene (M_2); neither of these monomers generates asymmetric centers in the copolymer chain. Conventional analysis of this system (Kinsinger et al., 1966) gives

$$k_{11}/k_{12} = r_1 = 3.3$$

$$k_{22}/k_{21} = r_2 = 0.05$$

These values indicate that vinylidene chloride radicals prefer to add vinylidene chloride, and that isobutylene radicals have only a very small tendency to add isobutylene. The system thus departs from any of the extreme cases so far considered. In Fig. 5.7 is the appropriate composition curve (Fig. 5.7b) together with those for four related systems for which $r_2 = 0$. Plots (a), (c), (d), and (e) start at $F_1 = 0.5$ when $f_1 = 0.0$; plot (b) begins at $F_1 = 0.0$ when $f_1 = 0.0$, since r_2 is small but nonzero; however, on the scale of Fig. 5.7 it too appears to begin at $F_1 = 0.5$.

The vinylidene chloride:isobutylene system has been studied by proton NMR (Hellwege et al., 1966; Kinsinger et al., 1966, 1967). In Fig. 5.8 are shown proton spectra of the homopolymers (a) and (b). The homopolymer of vinylidene chloride gives a single resonance for the methylene protons (a); the homopolymer of isobutylene, which can be prepared with cationic but not with free radical initiators, gives singlet resonances of 3:1 intensity for the methyl and methylene protons. The spectrum of a copolymer containing 70 mol% vinylidene chloride (m_1) is shown in (c). The resonances are grouped in three chemical shift ranges: m_1m_1-centered peaks at low field, the CH_2 and CH_3 resonances of m_2 units at high field, and peaks near 3 ppm that occur only in the copolymer spectra and must correspond to methylene protons of m_1m_2-centered units:

It is evident that tetrad sequences are resolved. The assignments are indicated in the caption. They are based partly on model compounds. If only dyad sequences were distinguished, there would be only three methylene resonances; for m_1m_1, $m_1m_2(m_2m_1)$, and m_2m_2 sequences. The upfield isobutylene peaks show considerable overlap and the assignments here are somewhat less certain. Also, the distinction between $m_2m_1m_2m_1$ and $m_1m_1m_2m_2$ sequences is not

$$m_2m_1m_2m_1$$

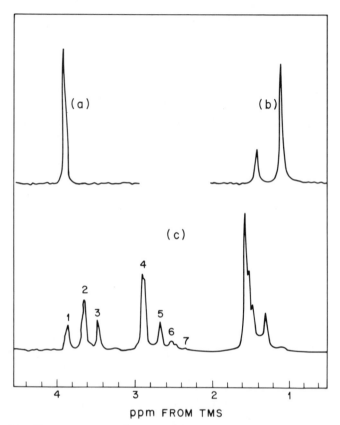

Fig. 5.8. The 60 MHz ^{1}H spectra of (a) poly(vinylidene chloride); (b) polyisobutylene; (c) a vinylidene chloride (m_1): isobutylene (m_2) copolymer containing 70 mol% m_1. Peaks are identified with monomer tetrad sequences (1) $m_1m_1m_1m_1$; (2) $m_1m_1m_1m_2$; (3) $m_2m_1m_1m_2$; (4) $m_1m_1m_2m_1$; (5) $m_2m_1m_2m_1$; (6) $m_1m_1m_2m_2$; (7) $m_2m_1m_2m_2$ (from Hellwege et al., 1966).

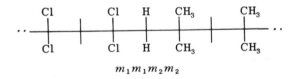

$$m_1m_1m_2m_2$$

entirely obvious; it is evident that the resonance of the latter exhibits fine structure, probably due to the resolution of hexads.

From dyad resonances only, r_1 and r_2 may in principle be calculated. The intensity of the resonances centered near 3.6 ppm gives $[m_1m_1]$ as 0.426 normalized over all dyad methylenes, which, upon substitution into Eq. (5.32), gives a value of 3.31 for r_1. To calculate r_2 it would be desirable to have a direct measure of $[m_2m_2]$, which is very small. This requires evaluation of the methylene resonance of

$$\cdots \begin{array}{ccc} CH_3 & H & CH_3 \\ | & | & | \\ \underset{\rule{0pt}{0pt}}{\overset{\rule{0pt}{0pt}}{}} & & \\ CH_3 & H & CH_3 \end{array} \cdots$$

$$m_2m_2$$

which occurs near 1.5 ppm, but cannot be readily separated from other resonances in this region, mainly methyl protons. An approximate value can be obtained from the relationship:

$$[m_1m_2](+[m_2m_1]) = 2F_2(1-P_{22}) \tag{5.40a}$$

or

$$[m_1m_2](+[m_2m_1]) = 2F_2 - \frac{2F_2 r_2 f_2}{1-f_2(1-r_2)} \tag{5.40b}$$

From the group of resonances near 2.8 ppm a value of $[m_1m_2]$ of 0.56 is obtained, from which a value for r_2 of approximately 0.04 is calculated. The reactivity ratios calculated from dyad intensities are thus in approximate agreement with those obtained by the conventional method. However, Kinsinger *et al.* (1967) showed that the proportions of m_1m_1-centered tetrads calculated from Eqs. (5.37)−(5.39) deviate slightly from the experimental values, while the proportions of the m_1m_2-centered tetrads—especially $m_1m_1m_2m_1$ and $m_2m_1m_2m_1$—deviate grossly. The terminal model thus does not strictly apply, although this is not a system for which "penultimate" effects, that is, an influence on reactivity of the monomer unit next to the terminal unit, would be expected to be marked. In such circumstances the mathematical treatment, while straightforward, becomes somewhat tedious. We must now consider eight rate constants and four reactivity ratios. Each monomer is represented by two reactivity ratios, one representing the propagating species in which the penultimate and terminal species are the same, and the other representing the propagating species in which they are different:

$$r_{11} = k_{111}/k_{112}, \qquad r_{22} = k_{222}/k_{221}$$

$$r_{21} = k_{211}/k_{212}, \quad r_{12} = k_{122}/k_{121}$$

Kinsinger *et al.* found that the expression of r_1 by a single value of 3.3 was too simple and that from tetrad analysis the relative reactivity of a growing free radical ending in vinylidene chloride depends on whether the penultimate unit is another vinylidene chloride or an isobutylene unit, so that

$$r_{11} = 2.95 \quad \text{and} \quad r_{21} = 6.22.$$

Thus, vinylidene chloride is more likely to add to a chain ending in vinylidene chloride if the unit next to the chain end is an isobutylene rather than another vinylidene chloride.

Figure 5.9 shows the proton spectrum of a vinylidene chloride:isobutylene copolymer very similar to that shown in Fig. 5.8 (with 65 mol% vinylidene chloride) but observed at 360 MHz, corresponding to a sixfold increase in separation of chemical shifts. The vinylidene chloride units are here identified as V and the isobutylene units as I. The resolution of sequences is now remarkably enhanced; the inset spectrum suggests that we can here resolve octad sequences. (Some of the assignments to longer sequences are conjectural.) Clearly, very searching tests of mechanistic proposals can be made with such information. A point of some interest can be seen in the 1.1 ppm region of the spectrum. Here, one can observe the methyl resonances of isobutylene sequences, and it is observed that not only $m_2 m_2$ dyads but $m_2 m_2 m_2$ triads (III) can be resolved. Analysis of this part of the spectrum would yield accurate values of r_2 (or r_{22} and r_{12}) but this has not yet been done.

Let us now consider a case in which stereochemical configuration makes a marked contribution to the NMR spectrum. One of the earliest systems to be studied by NMR was the free radical copolymerization of styrene and methyl methacrylate (Bovey, 1962; Harwood, 1965; Harwood and Ritchey, 1965; Ito and Yamashita, 1968; Nishioka *et al.*, 1962; Kato *et al.*, 1964; Overberger and Yamamoto, 1965; Ito and Yamashita, 1965; Bauer *et al.*, 1966; Ito *et al.*, 1967; Yabumoto *et al.*, 1970; Katritzky *et al.*, 1974; Katritzky and Weiss, 1976; Koinuma *et al.*, 1980). In Table 5.2 are shown the structures, through triad sequences, which must be considered in such copolymer chains, excluding inversions, branches, and end-groups (Chapter 6). The representation of longer sequences is omitted, for we find that the number of sequences $N'(n)$ increases very rapidly with n:

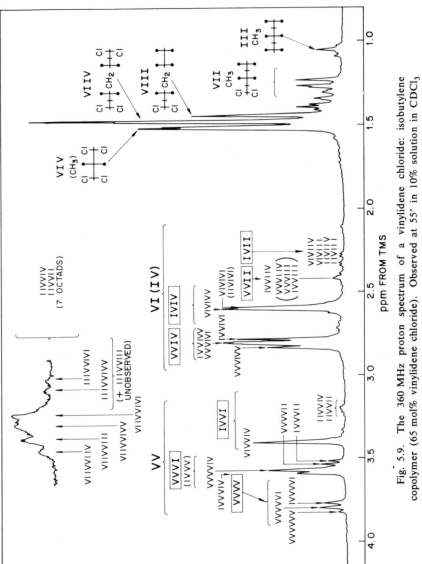

Fig. 5.9. The 360 MHz proton spectrum of a vinylidene chloride: isobutylene copolymer (65 mol% vinylidene chloride). Observed at 55° in 10% solution in $CDCl_3$ (R. E. Cais, private communication, 1975).

n	2	3	4	5	6
$N'(n)$	6	20	72	272	1056

or

$$N'(n) = 2^{2(n-1)} + 2^{n-1}$$

If monomer 1 is taken as styrene and monomer 2 as methyl methacrylate, we find (Table 5.1)

$$k_{11}/k_{12} = r_1 = 0.52$$

TABLE 5.2

Configurational Sequences in Copolymers

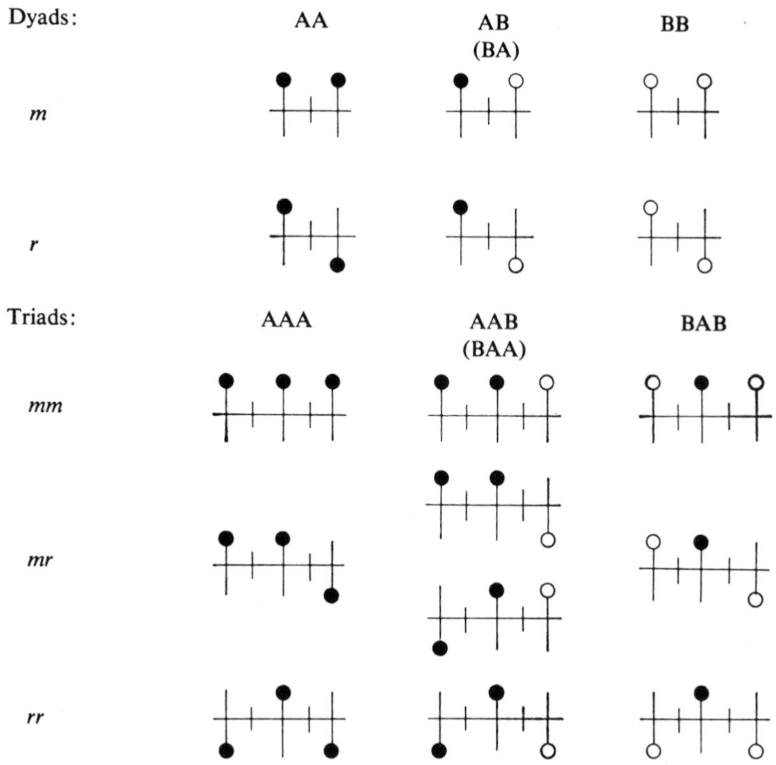

Dyads:	AA	AB (BA)	BB
m			
r			

Triads:	AAA	AAB (BAA)	BAB
mm			
mr			
rr			

+ 10 others with ● and ○ reversed.

$$k_{22}/k_{21} = r_2 = 0.46$$

$$r_1 r_2 = 0.24$$

Thus, each growing chain prefers to add the other monomer by a factor of about 2, and the system is of the crossover or alternating type. It corresponds quite closely to the 0.5/0.5 plot in Fig. 5.4. In Fig. 5.10 are shown 40 MHz proton spectra of a series of styrene—methyl methacrylate copolymers ranging in feed composition from polystyrene (a) to poly(methyl methacrylate) (g); in (b) $f_1 = 0.90$;   (c) $f_1 = 0.75$;   (d) $f_1 = 0.50$;   (e) $f_1 = 0.25$; (f) $f_1 = 0.10$. (It is noteworthy that the observation of these copolymers at frequencies higher than 40 MHz reveals little further detail, no doubt because the many possible sequences form broad unresolvable resonances; however, this is not the case for all copolymers, as we have seen.) It would be possible in principle but not easy in fact to determine r_1 and r_2 from such data. What is actually revealed most prominently is the stereochemistry of the copolymers, which is scarcely recognized as existing in most conventional studies of copolymers, although it clearly must affect physical properties, if less so than in homopolymers. Somewhat surprisingly, the methoxyl proton resonance, which is quite insensitive to stereochemistry in poly(methyl methacrylate), is very sensitive to the copolymer stereochemistry. We see that this resonance (at 2.0−3.4 ppm) is split into three to five peaks. Even when methyl methacrylate is only a minor fraction of the copolymer, as in (b), where F_2 is ca. 0.12, the methoxyl resonance is clearly split. This cannot be due to comonomer sequences, for $m_1 m_2 m_1$ predominates very strongly (over 98%), the probability of $m_1 m_2 m_2$ and $m_2 m_2 m_2$ being very low. The most probable assignments are given in Table 5.3 (representing monomer triads and their configurations in planar zigzag projection); mixed dyads are termed "co-meso" and "co-racemic", abbreviated co-m and co-r, respectively. Thus, co-racemic placements of styrene on both sides of the methyl methacrylate unit leave the methoxyl resonance almost where it is in the homopolymer. One co-meso styrene unit shields it by ca. 0.5 ppm; two co-meso units shield it by an additional 0.5 ppm. This shielding effect is to be attributed to the phenyl group ring current.

At $f_1 = 0.50$, corresponding to approximately $F_1 = 0.50$ [spectrum (d)], comonomer sequence effects begin to be evident, as there are now five resolved methoxyl resonances. Of these, $m_1 m_2 m_1$

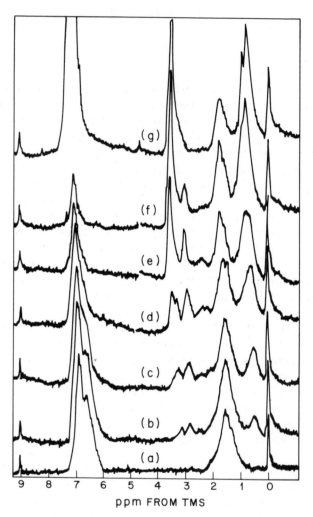

Fig. 5.10. The 40 MHz proton spectra of methyl methacrylate: styrene copolymers and homopolymers (0.10 g in 0.5 ml of $CHCl_3$ or CCl_4 ; 120°): (a) polystyrene; (b) 10:90 (methyl methacrylate: styrene mole ratio in monomer feed), i.e., $f_1 = 0.90$; (c) 25:75, $f_1 = 0.75$; (d) 50:50, $f_1 = 0.50$; (e) 75:25, $f_1 = 0.25$; (f) 90:10, $f_1 = 0.10$; (g) poly(methyl methacrylate).

TABLE 5.3

¹H NMR Assignments for Methyl Methacrylate: Styrene Copolymers

Sequence	Designation	OCH_3 Proton Chemical Shift
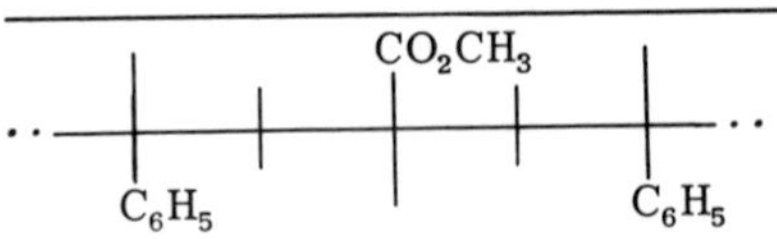 [structure: CO_2CH_3 with $C_6H_5 \cdots C_6H_5$]	$co\text{-}r,\ co\text{-}r$	3.3 ppm
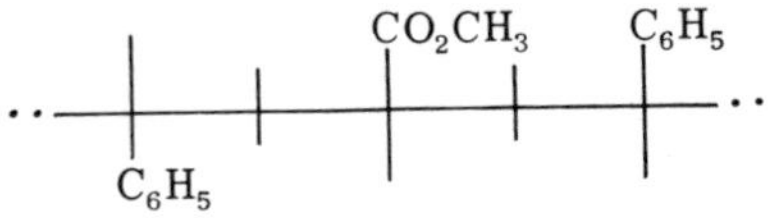 [structure: CO_2CH_3, C_6H_5 with C_6H_5]	$co\text{-}r,\ co\text{-}m$	2.8 ppm
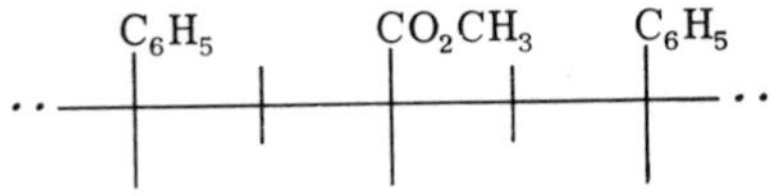 [structure: C_6H_5, CO_2CH_3, C_6H_5]	$co\text{-}m,\ co\text{-}m$	2.3 ppm

and $m_2 m_2 m_1$ must each constitute a fraction equal to 0.45, while $m_2 m_2 m_2$ is 0.10. Therefore, the least shielded methoxyl resonances are probably

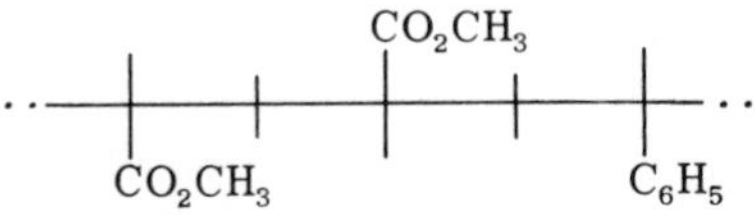

[structure: CO_2CH_3 ... CO_2CH_3 ... C_6H_5] and [structure: CO_2CH_3 CO_2CH_3 ... C_6H_5]

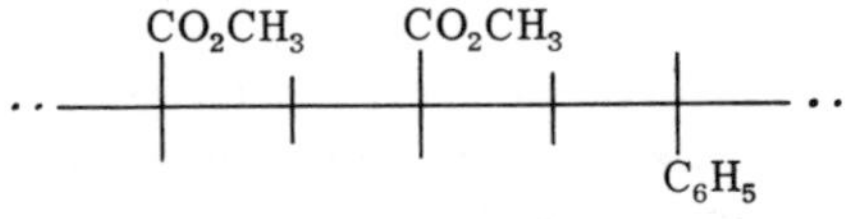

appearing at *ca.* 3.5 ppm, while the slightly more shielded resonance at *ca.* 3.4 ppm may be

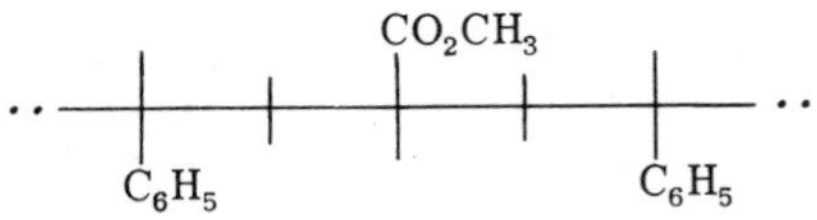

[structure: CO_2CH_3 with C_6H_5 ... C_6H_5]

The more shielded resonances are assigned as before.

The ^{13}C spectra of methyl methacrylate—styrene copolymers show better resolution of comonomer and stereochemical sequences (Katritzky *et al.*, 1974). The α—CH$_3$ resonances of the methyl methacrylate units and the aromatic C-1 carbons of the styrene units

are the most sensitive to tacticity. It is concluded that methyl methacrylate—styrene sequences are nearly random in configuration, while runs or blocks of each monomer are predominantly syndiotactic. This second conclusion is entirely reasonable on the basis of what we have learned concerning the stereochemical configuration of the homopolymers.

It is to be expected that copolymers having long sequences of styrene and methyl methacrylate units will give spectra that approximate to summations of the homopolymer spectra, with little or no complexity in the methoxyl region. This is indeed found for anionic copolymers (Overberger and Yamamoto, 1965; Ito and Yamashita, 1965). Of greater interest are copolymers formed from complexed monomers. Hirai *et al.* (1979; Koinuma *et al.*, 1980) found that the copolymer of styrene and methyl methacrylate prepared photochemically in the presence of stannic chloride or ethylaluminum sesquichloride as complexing agents gave spectra quite different from that of the normal free radical product. It is known from earlier studies that such copolymers have a 50:50 alternating structure. In Fig. 5.11 is shown the 100 MHz proton spectrum of a copolymer with α-d_1-styrene, observed in pentachloroethane at 160°. The spectral features are more clearly resolved than in Fig. 5.10. This may be in part due to the higher observing frequency, but mainly arises from the greater structural simplicity of the copolymer chains, in which only three types of methyl methacrylate-centered triads can occur:

co-syndio   co-hetero

$$C_6H_5 \qquad CO_2CH_3 \qquad C_6H_5$$

$$CH_3$$

co-iso

It will be noted that the assignments of the methoxyl resonances are in accord with the previous discussion, but that the $\alpha-$methyl resonances are assigned in reverse order. This appears logical, based on the planar zigzag projections, i.e., trans conformations, but Koinuma *et al.* were able to support it with more sophisticated calculations based on a rotational isomeric state model (see Chapter 7).

Many other copolymer systems have been subject to scrutiny by ^{1}H or ^{13}C NMR, and a vast literature has accumulated. We shall close this discussion by briefly considering a system of much intrinsic interest, which is also the basis of an important class of synthetic elastomers: that of ethylene−propylene. This comonomer pair has been extensively studied by ^{13}C spectroscopy (Crain *et al.*, 1971; Zambelli *et al.*, 1971; Carman and Wilkes, 1971; Carman *et al.*, 1977; Randall, 1977). It presents complications that we have not yet considered. First, there is a readily observable content of inverted or

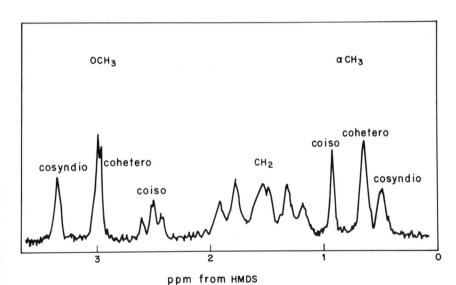

Fig. 5.11. The 100 MHz proton spectrum of a 1:1 alternating copolymer of methyl methacrylate and styrene, observed in pentachloroethane at 160° (Koinuma *et al.*, 1980).

head-to-head:tail-to-tail propylene units. These are best treated as a third comonomer (Chapter 6). In addition, the copolymerization cannot always be assumed to obey the assumptions inherent in the copolymer equation. With insoluble coordination catalysts such as VCl_3, together with aluminum alkyls or chloroaluminum alkyls, there may be two or more catalytic sites producing different polymer chains—some rich in ethylene and others rich in propylene—and such a system clearly cannot conform to any simple mechanistic picture. However, with soluble vanadium compounds (e.g., $VOCl_3$, VCl_4, or vanadium triacetylacetonate) it is believed (Crespi et al., 1977) that the reaction takes place in a homogeneous phase and can be treated as a fairly normal system. There is some difference in reported reactivity ratios depending on the particular catalyst employed, but most of these differences probably represent experimental error in determining these ratios.

The reactivity ratios determined by classical means (see Crespi et al., 1977) clearly show that ethylene (monomer 1) is much more reactive than propylene: r_1 values are in the 16 ± 2 range, whereas r_2 is of the order of 0.05. The reported values of r_1r_2 are close to 1, or somewhat lower, indicating (p. 128) that the nature of the chain end has little influence on the relative rates of addition of the monomers. Figure 5.12 shows the 25 MHz ^{13}C spectrum of a 50:50 (mole ratio) ethylene–propylene copolymer, observed in 1,2,4-trichlorobenzene at 120°C (Randall, 1977); it is very similar to a spectrum reported by Carman et al. (1977), but shows notably better resolution. The peaks were initially assigned by Crain et al. (1977) and Zambelli et al. (1971) on the basis of the Grant and Paul rules for carbon chemical shifts (which we shall briefly discuss in Section 7.8). Later, more detailed assignments were given by Carman (Carman and Wilkes, 1971; Carman et al., 1977). The principal chemical shift influence is that of the methyl-branched carbons, but stereochemical configuration is a substantial perturbation, as we might expect from the discussion of polypropylene (Section 3.6.5).

An early analysis by Carman and Wilkes (1971) relied entirely on the methine resonances; since these are relatively insensitive to both configuration and monomer inversion, this restriction simplifies the interpretation but has the disadvantage that at least 80% of the spectral information is ignored. Methyl resonances are very sensitive to configuration as well as monomer sequence, and this complicates their assignment. The result, as even a superficial examination of Fig. 5.12 might lead one to expect, is that the methylene carbon resonances are

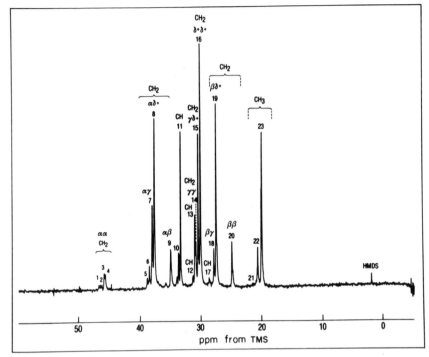

Fig. 5.12. The 25 MHz ^{13}C spectrum of an ethylene–propylene copolymer of 50:50 mole ratio, observed at 120°C in 1,2,4-trichlorobenzene (Randall, 1977).

the most useful. In a sequence of methylene carbons between branch carbons, each carbon gives a distinctive resonance up to a run length of 7. The assignments established by Carman and Wilkes are shown below. The Greek letter designations are those employed by them,

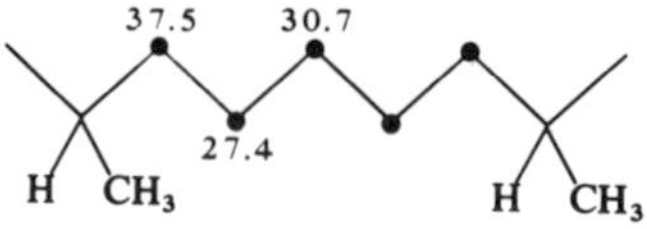

$\alpha\delta^+, \beta\delta^+, \gamma\gamma, \beta\delta^+, \alpha\delta^+$

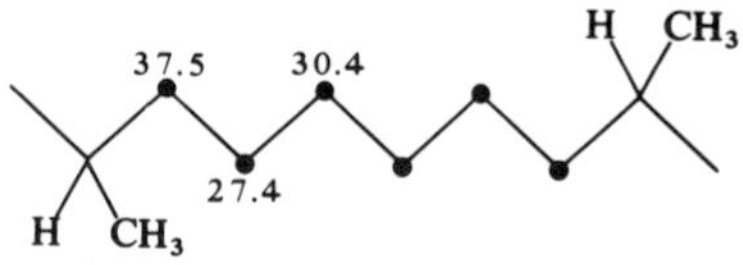

$\alpha\delta^+, \beta\delta^+, \gamma\delta^+, \gamma\delta^+, \beta\delta^+, \alpha\delta^+$

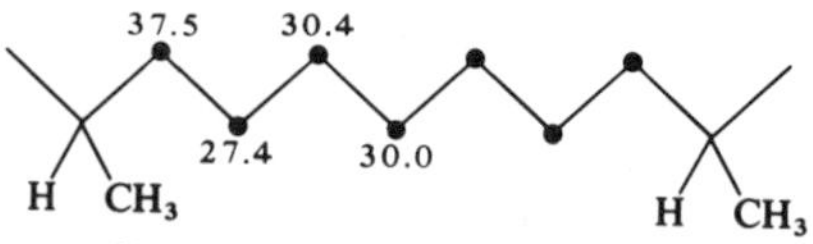

$\alpha\delta^+, \beta\delta^+, \gamma\delta^+, \delta^+\delta^+, \gamma\delta^+, \beta\delta^+, \alpha\delta^+$

and indicate the position of the nearest branch carbon. A plus sign indicates that the branch is at the δ or more distant position. The methylene carbons that are four or more carbons removed from a branch carbon give the same chemical shift (30.0 ppm) at 25 MHz; it can be anticipated that at higher observing frequencies finer discrimination will be possible. Propylene inversions can be clearly identified through the presence of methylene sequences two and four carbons in length.

From these assignments and a mathematical model of the chain growth that treats inverted propylene units as monomer 3, Carman and Wilkes completely analyzed the statistical distribution of sequences in three copolymers containing 23.7, 43.1, and 53.2 mol% of propylene. There are nine propagation probabilities: $P_{11}, P_{12}, P_{13}, P_{21}, P_{22}, P_{23}, P_{31}, P_{32}$, and P_{33}, but these are reduced to five because of necessary relationships between them and because one, P_{32}, corresponding to the head-to-head dyad

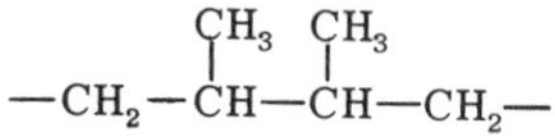

$$-CH_2-\overset{\overset{\displaystyle CH_3}{\mid}}{CH}-\overset{\overset{\displaystyle CH_3}{\mid}}{CH}-CH_2-$$

can be set to zero, as no resonances corresponding to this structure can be detected in the spectra. The extent of propylene inversion was of the order of 5 mol%, but the estimated error was very large. A value of $r_1 r_2$ of 0.35 ± 0.04 was deduced. The analysis was long and rather complicated. The reader is referred to the original paper (Carman *et al.*, 1977) for details and for an illuminating discussion.

REFERENCES

Bauer, R. G., Harwood, H. J., and Ritchey, W. M. (1966). *Polymer Preprints* **7** (2), 973.
Bovey, F. A. (1962). *J. Polymer Sci.* **62**, 197.
Bowden, M. J. (1979). *In* "Macromolecules" (F. A. Bovey and F. H. Winslow, eds.), pp. 138-139. Academic Press, New York.
Carman, C. J., and Wilkes, C. E. (1971). *Rubber Chem. Technol.* **44**, 781.
Carman, C. J., Harrington, R. A., and Wilkes, C. E. (1977). *Macromolecules* **10**, 536.
Crain, W. O., Jr., Zambelli, A., and Roberts, J. D. (1971). *Macromolecules* **4**, 330.
Crespi, G., Valvassori, A., and Flisi, U. (1977). *In* "The Stereo Rubbers" (W. M. Saltman, ed.), p. 365. Wiley, New York.
Elias, H. G. (1977). "Macromolecules," p. 768. Plenum, New York.
Fineman, M., and Ross, S. D. (1950). *J. Polymer Sci.* **5**, 259.
Goldfinger, G., and Kane, T. (1948). *J. Polymer Sci.* **3**, 462.
Gumbs, R., Penczek, S., Jagur-Grodzinski, J., and Szwarc, M. (1969). *Macromolecules* **2**, 77.
Harwood, H. J. (1965). *Angew. Chem. Int. Ed. Eng.* **4**, 1051.
Harwood, H. J., and Ritchey, W. M. (1965). *J. Polymer Sci., Part B* **3**, 419.
Hellwege, K. H., Johnsen, U., and Kolbe, K. (1966). *Kolloid-Z.* **214**, 45.
Hirai, H., Tanabe, T., and Koinuma, H. (1979). *J. Polymer Sci. Polymer Chem. Ed.* **17**, 843.
Ito, K., and Yamashita, Y. (1965). *J. Polymer Sci., Part B* **3**, 625, 631.
Ito, K., and Yamashita Y. (1968). *J. Polymer Sci., Part B* **6**, 227.
Ito, K., Iwase, S., Umehara, K., and Yamashita, Y. (1967). *J. Macromol. Sci., Part A* **1**, 891.
Kato, Y., Ashikari, N., and Nishioka, A. (1964). *Bull. Chem. Soc. Japan* **37**, 1630.
Katritzky, A. R., and Weiss, D. E. (1976). *Chem. Britain*, 45.
Katritzky, A. R., Smith, A., and Weiss, D. E. (1974). *J. Chem. Soc. Perkin II* **1974** 1547.
Kinsinger, J. B., Fischer, T., and Wilson, C. W. III (1966). *J. Polymer Sci., Part B* **4**, 379.
Kinsinger, J. B., Fischer, T., and Wilson, C. W. III (1967). *J. Polymer Sci., Part B* **5**, 285.
Koinuma, H., Tanabe, T., and Hirai, H. (1980). *Makromol. Chem.* **181**, 383.
Mayo, F. R., and Lewis, F. M. (1944). *J. Am. Chem. Soc.* **66**, 1594.
Melville, H. W., Noble, B., and Watson, W. F. (1947). *J. Polymer Sci.* **2**, 229.
Nishioka, A., Kato, Y., and Ashikari, N. (1962). *J. Polymer Sci.* **62** S 10.
Overberger, C. G., and Yamamoto, N. (1965). *J. Polymer Sci., Part B* **3**, 569.
Randall, J. C. (1977). "Polymer Sequence Determination, Carbon-13 NMR Method," pp. 53, 135. Academic Press, New York.
Yabumoto, S., Ishii, K., and Arita, K. (1970). *J. Polymer Sci., Part A-1* **8**, 295.
Zambelli, A., Gatti, G., Sacchi, C., Crain, W. O., Jr., and Roberts, J. D. (1971). *Macromolecules* **4**, 475.

Chapter 6

REGIOREGULARITY AND BRANCHING IN VINYL POLYMER CHAINS

6.1 INTRODUCTION

In this chapter we discuss in some detail two additional forms of isomerism in vinyl polymer chains that were only briefly noted in Chapter 1. By *regioregularity* (or lack of it) we refer to head-to-tail versus head-to-head:tail-to-tail isomerism. If the propagation reaction produces one or the other of these exclusively, or nearly so, it is described as *regiospecific*. This term is borrowed from the organic chemist, who employs it with reference to positional isomerism in the direction of addition to a double bond or in the substitution of an aromatic ring. It could also logically be employed in polymer chemistry to describe the mode of propagation of diene monomers, i.e., in the generation of geometrical isomers (Chapter 4) but we shall not use it in this sense in this book.

Branching (and crosslinking) may also be regarded as forms of isomerism in vinyl and diene polymer chains. We shall consider their observation principally in polyethylene and poly(vinyl chloride), which have been studied the most intensively in this regard.

157

6.2 HEAD-TO-TAIL VERSUS HEAD-TO-HEAD:TAIL-TO-TAIL ISOMERISM; REGIOREGULARITY

Vinyl polymers are usually predominantly head-to-tail or *isoregic*[*] (i):

$$
\cdots -\underset{\underset{B}{|}}{\overset{\overset{A}{|}}{C}}-\underset{\underset{H}{|}}{\overset{\overset{H}{|}}{C}}-\underset{\underset{B}{|}}{\overset{\overset{A}{|}}{C}}-\underset{\underset{H}{|}}{\overset{\overset{H}{|}}{C}}-\underset{\underset{B}{|}}{\overset{\overset{A}{|}}{C}}-\underset{\underset{H}{|}}{\overset{\overset{H}{|}}{C}}-\underset{\underset{B}{|}}{\overset{\overset{A}{|}}{C}}-\underset{\underset{H}{|}}{\overset{\overset{H}{|}}{C}}- \cdots
$$

(i)

Polymers may also be of head-to-head:tail-to-tail or *syndioregic*[*] structure (ii):

$$
\cdots -\underset{\underset{H}{|}}{\overset{\overset{H}{|}}{C}}-\underset{\underset{B}{|}}{\overset{\overset{A}{|}}{C}}-\underset{\underset{B}{|}}{\overset{\overset{A}{|}}{C}}-\underset{\underset{H}{|}}{\overset{\overset{H}{|}}{C}}-\underset{\underset{H}{|}}{\overset{\overset{H}{|}}{C}}-\underset{\underset{B}{|}}{\overset{\overset{A}{|}}{C}}-\underset{\underset{B}{|}}{\overset{\overset{A}{|}}{C}}-\underset{\underset{H}{|}}{\overset{\overset{H}{|}}{C}}- \cdots
$$

(ii)

Pure syndioregic chains must be prepared by indirect routes. For example, syndioregic polystyrene can be prepared by the free radical 1,4-polymerization of 2,3-diphenylbutadiene, followed by reduction (Vogel, 1979):

$$
CH_2 = \underset{\overset{|}{C_6H_5}}{\overset{\overset{H_5C_6}{|}}{C}}-\underset{\overset{|}{C_6H_5}}{C} = CH_2 \longrightarrow \left[CH_2 - \underset{\overset{|}{C_6H_5}}{\overset{\overset{H_5C_6}{|}}{C}} = \underset{\overset{|}{C_6H_5}}{C} - CH_2 \right]_n \longrightarrow
$$

$$
\left[CH_2 - \underset{\overset{|}{C_6H_5}}{\overset{\overset{H_5C_6}{|}}{CH}} - \underset{\overset{|}{C_6H_5}}{CH} - CH_2 \right]_n
$$

The representation of syndioregic chains in projection requires a word of explanation. It may be recalled (Chapter 3, Appendix) that isoregic vinyl polymer chains appear the same in either Fischer or planar zigzag projection, but this is not true for polymers of 1,2-

---

[*] I am indebted for this nomenclature to Dr. R. E. Cais. It is based on the same Greek roots *iso* ("equal") and *syndio* ("every two") used by Natta in devising the terms employed for stereochemical configuration.

disubstituted monomers. Syndioregic chains are similar in this respect to the latter. Thus, the unit

$$\cdots -CH_2- \overset{\overset{\displaystyle R}{|*}}{CH} - \overset{\overset{\displaystyle R}{|*}}{CH} - CH_2- \cdots$$

may be shown as

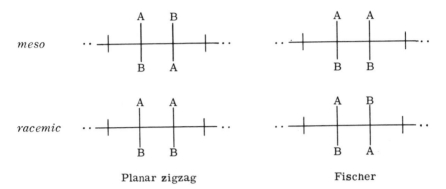

Planar zigzag   Fischer

We shall employ planar zigzag projections in this discussion. It should also be noted that the asterisked carbons are in this case true asymmetric centers; the meso unit could be represented as *RS* (or *SR*) and the racemic unit as *RR* (or *SS*), although it is very unlikely that the absolute chiralities will be known in any real case.

It is evident that higher orders of configurational isomerism can be recognized and may be manifested spectroscopically. For example, if our concern is with the methylene groups (or fluoromethylene groups) joining the dyads of asymmetric centers, we can recognize

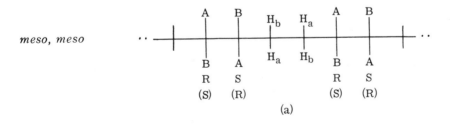

(a)

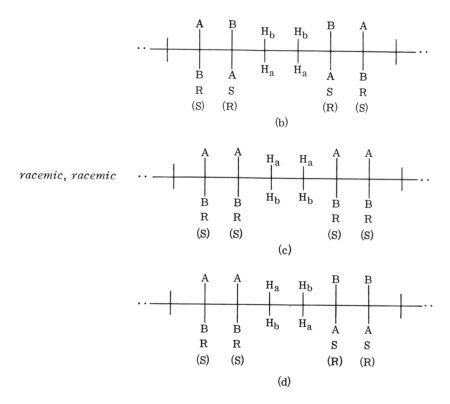

(b)

racemic, racemic

(c)

(d)

With respect to the asymmetric carbons and the substituents, one can recognize triads of dyads, e.g.:

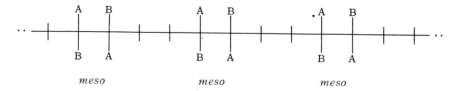

and many other sequences that may be readily written down.

In vinyl monomer propagation the occurrence of inversions is, as we have said, comparatively infrequent, usually taking the form of single inverted units:

$$\cdots -CH_2-\underset{B}{\overset{A}{C}}-CH_2-\underset{B}{\overset{A}{C}}-\underset{B}{\overset{A}{C}}-CH_2-CH_2-\underset{B}{\overset{A}{C}}-CH_2-\underset{B}{\overset{A}{C}}-\cdots$$

In free radical polymerization chain termination may occur by recombination, producing a head-to-head unit with no corresponding tail-to-tail junction:

$$\cdots - CH_2 - \underset{\underset{B}{|}}{\overset{\overset{A}{|}}{C}} - CH_2 - \underset{\underset{B}{|}}{\overset{\overset{A}{|}}{C}} - \underset{\underset{B}{|}}{\overset{\overset{A}{|}}{C}} - CH_2 - \underset{\underset{B}{|}}{\overset{\overset{A}{|}}{C}} - CH_2 - \cdots$$

These can be directly detected by ^{13}C NMR in poly(methyl methacrylate) (F. C. Schilling, private communication, 1981).

6.2.1 Poly(vinyl chloride)

In the earlier days of polymer science it was not known whether vinyl polymers had isoregic or syndioregic structures. Marvel and his co-workers (1939) undertook a study of the structure of poly(vinyl chloride) employing the Freund reaction with metallic zinc, which was believed to be specific for the removal of chlorine atoms in 1,3 positions:

$$\underset{\underset{Cl}{|}}{CH} \quad \underset{\underset{Cl}{|}}{CH} \quad + \quad Zn \quad \longrightarrow \quad CH - CH \quad + \quad ZnCl_2$$

Thus, if the chains were syndioregic the reaction would not be expected to proceed. Even if they were isoregic the reaction would not be complete if it had the assumed specificity. If the reaction is random and irreversible it will generate some isolated chlorine atoms, which presumably cannot react since the zinc must remove chlorines in pairs:

$$\underset{\underset{Cl}{|}}{CH} \ \underset{\underset{Cl}{|}}{CH} \ \underset{\underset{Cl}{|}}{CH} \ \underset{\underset{Cl}{|}}{CH} \ \underset{\underset{Cl}{|}}{CH} \longrightarrow CH - CH \ \ \underset{\underset{Cl}{|}}{CH} \ \ CH - CH$$

It was shown by Flory (1939) that the limiting fraction of chlorine that becomes unremovable is e^{-2} or 13.53%. Marvel *et al.* (1939) reported that a maximum of 84−87% could be removed, apparently confirming expectation. (In their work there was no direct confirmation of the formation of cyclopane rings, since this was virtually impossible in prespectroscopic days.) However, closer examination raised some questions. The formation of double bonds was detected (Alfrey *et al.*, 1951, 1952a,b; Smets, 1966) A 60 MHz proton NMR study

(Tepelekian *et al.*, 1969) showed the presence of cyclopropane rings but also revealed more methylene groups than expected.

Using proton and ^{13}C NMR, Cais and Spencer (1982) made a searching examination of the structure of zinc dechlorinated poly(vinyl chloride), which was also further reduced to the hydrocarbon with tri-*n*-butyltin hydride, as described in detail in Section 6.3.2. The latter reagent removes chlorine regardless of its placement, including "lone" chlorine atoms, thereby simplifying the spectrum, although with some loss of information. We describe this study as an example of the power of modern spectroscopy to unravel a complicated microstructural problem. Dechlorination was carried out with zinc dust in dioxane with zinc chloride catalyst; the reaction was carried to 4, 56, 61, and 79% of complete removal. In Fig. 6.1 is shown the proton-decoupled 90.5 MHz ^{13}C spectrum of the polymer at 56% removal. Its complexity is a faithful reflection of the structural complexity of the chain. From the proton-coupled spectrum and from model compounds, most of these resonances can be assigned.

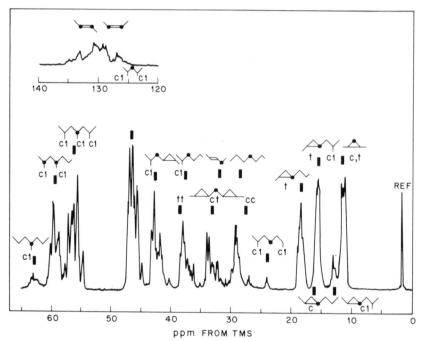

Fig. 6.1. The 90.5 MHz ^{13}C NMR spectrum of poly(vinyl chloride) after removal of 56% of chlorine by zinc. Observed in 1,2,4-trichlorobenzene at 90° (Cais and Spencer, 1982).

The resonances centered at 130 ppm correspond to vinylene (cis and trans) carbons in a variety of local environments. The 10−65 ppm region corresponds to saturated carbons, divisible into 11 groups of peaks. Their assignments are as shown, with the observed carbon indicated by a black dot. The fine structure is due to the effects of longer sequences and configurational isomerism. Least shielded are the carbons at *ca.* 63 ppm, which bear a single chlorine substituent. Chlorine atoms in a γ position, i.e., three bonds distant, cause additional shielding owing to the γ-*gauche* effect, which we shall discuss further in Chapter 7.

Acyclic methylene carbons experience a variety of environments and exhibit shieldings in the 25−50 ppm range. They are most deshielded by chlorine atoms on neighboring carbons, less so by adjacent double bonds or cyclopropyl groups. The center methylene carbon in a sequence of five (or more) methylenes resonates at *ca.* 30.0 ppm. Methylenes in cyclopropane rings are the most shielded, appearing at *ca.* 12 ppm. Methylenes between cyclopropyl groups (29−38 ppm) are highly sensitive to the geometry of the neighboring rings, being least shielded when these are trans and most shielded when they are cis.

Figure 6.2 shows the simplification of the ^{13}C spectrum resulting from reductive dehalogenation with tributyltin hydride. It now can be clearly seen that the cyclopropyl rings are over 70% trans (compare the peaks at 29.1 and 34.6). On a simple ball-and-stick basis it would be expected that meso dyads would give cis rings and racemic dyads trans rings. However, the zinc reduction and ring closure is a free radical reaction, and the cis:trans ratio probably represents an inherent preference in the transition state rather than the ratio of starting structures (the initial polymer is about 57% racemic). The cyclopentane rings are believed to arise not from 1,5 chlorine removal by zinc but as an artifact in the tributyltin hydride reduction.

It is evident that the Freund reaction takes the textbook course only in part. It was found that the chlorine removal leveled off at about 79% of complete removal, but the ^{13}C spectrum (not shown) of the polymer at this final stage shows that half the chlorine is still in runs of three or longer, an observation that does not appear compatible with random pairwise removal. In addition, 9% of the original -CH$_2$CHCl- units have been transformed into -CH=CH- units and 21% have been reduced to -CH$_2$CH$_2$- units (the latter probably arising by free radical hydrogen removal from the dioxane solvent). Thus, of every ten -CH$_2$CHCl- units ideally expected to form cyclopropane rings, only six actually do so.

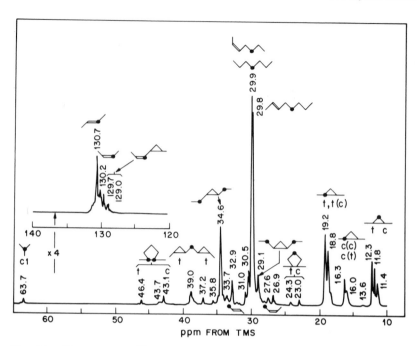

Fig. 6.2.   The 90.5 MHz ^{13}C NMR spectrum of poly(vinyl chloride) after removal of 79% of chlorine by zinc and complete reductive dechlorination with tri(n-butyl)tin hydride. Observed in 1,2,4-trichlorobenzene at 90° (Cais and Spencer, 1982).

The Freund reaction as a means of establishing an isoregic structure for poly(vinyl chloride) is thus very seriously flawed. Nevertheless, the conclusion of Marvel *et al.* is correct. This can be demonstrated more directly by the observation of poly(vinyl chloride) prepared using ^{13}C enrichment at both the $\alpha-$ and $\beta-$carbons (W. H. Starnes, Jr. and F. C. Schilling, private communication, 1981), the spectrum of which shows none of the resonances to be expected of inverted units.

6.2.2 Poly(vinyl alcohol)

A considerably more subtle but still prespectroscopic approach to the measurement of head-to-head units is that applied by Flory and Leutner (1948, 1950) to poly(vinyl alcohol). It again is of very limited application and depends on selective chemical reactivity. It was by this time known that the predominant structure is isoregic and the question now posed was the frequency of occurrence of the occasional inverted unit. It was assumed that periodic acid reacts only with 1,2-glycol units and not with 1,3 units:

$$\cdots - CH_2 - \underset{\underset{OH}{|}}{CH} - CH_2 - \underset{\underset{OH}{|}}{CH} - \underset{\underset{OH}{|}}{CH} - CH_2 - CH_2 - \underset{\underset{OH}{|}}{CH} - \cdots$$

$$\xrightarrow{HIO_4} \cdots - CH_2CHO + HOCCH_2 - \cdots$$

Inverted units in poly(vinyl alcohol) thus lead to chain scission, the extent of which was measured by the reduction of solution viscosity. This is an inherently quite sensitive method. It was concluded that 1–2% of head-to-head units are present, this mode of propagation in the original free radical vinyl acetate polymerization increasing with temperature to a degree corresponding to an additional energy of activation of *ca.* 1.3 kcal-mol^{-1}. The method depends of course on very sharp discrimination by the periodic acid. If the 1,3-glycol units react even 1% as fast as the 1,2-units, there will be a twofold error.

6.2.3 Fluorovinyl Polymers

Fluorine-substituted ethylenes are particularly subject to the generation of regioirregular chains, presumably because fluorine atoms are relatively undemanding sterically. At the same time, the physical properties of their polymers may depend strongly, even critically, on the presence of inverted units. The presence of ^{19}F offers an additional and powerful means for detailed study, as ^{19}F chemical shifts are highly sensitive to structural variables.

6.2.3.1 Poly(vinyl fluoride)

Poly(vinyl fluoride) is produced on a limited commercial scale for specialized uses; it is a highly weather-resistant plastic, optically clear despite a high degree of crystallinity. Its structure is quite irregular, as first demonstrated by Wilson and Santee (1965) by ^{19}F NMR. They estimated that a substantial fraction of the monomer units, perhaps as many as one out of six, were inverted. However, assignments of fluorine resonances were somewhat conjectural. In Fig. 6.3, spectrum (a) is the 188.2 MHz ^{19}F spectrum of a commercial material. The $^{19}F - {}^{1}H$ coupling multiplicity has been removed by proton irradiation. The resonances, designated by capital letters, are assigned according to the following scheme:

$$\cdots - CH_2 \overset{A}{\underset{\longrightarrow}{-}} CHF - CH_2 \overset{A}{\underset{\longrightarrow}{-}} CHF - CH_2 \overset{B}{\underset{\longrightarrow}{-}} CHF \overset{C}{\underset{\longleftarrow}{-}} CHF - CH_2 - CH_2 \overset{D}{\underset{\longrightarrow}{-}} CHF - CH_2 \overset{A}{\underset{\longrightarrow}{-}} CHF - \cdots$$

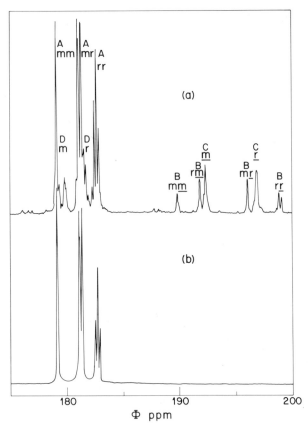

Fig. 6.3. The 188.2 MHz ^{19}F spectrum of (a) commercial poly(vinyl fluoride); (b) poly(vinyl fluoride) prepared by reductive dechlorination of poly(1-fluoro-1-chloroethylene); both spectra observed at 130°C in 8% (w/v) solution in *N,N*-dimethylformamide (Tonelli *et al.*, 1982; R. E. Cais and J. Kometani, private communication,1982).

The stereochemical assignments are also indicated, and require a further word of explanation. The *m* and *r* designations that are not underlined represent the usual relationships between substituents in 1,3 positions (in planar zigzag projection),

whereas the underlined designations represent the substituents in 1,2 positions (also in planar zigzag projection):

$$m \qquad\qquad r$$

(It may further be noted in connection with the resonances of B that $\underline{rm}$ and $\underline{mr}$ are quite different structures, and do not differ merely in direction as with rm and mr.)

Spectrum (b) is that of a poly(vinyl fluoride) prepared by the following route:

$$CH_2{=}CFCl \xrightarrow[\text{initiator}]{\substack{\text{free} \\ \text{radical}}} {+}CH_2{-}CFCl{+}_n \xrightarrow{Bu_3SnH} {+}CH_2{-}CHF{+}_n$$

The steric requirements of the chlorine atom permit only a negligible proportion of head-to-head units, and it is now observed that the upfield portion of the spectrum, as well as resonances D, are absent.

It is evident from spectrum (a) that poly(vinyl fluoride), as normally prepared, is nearly atactic; peak intensity measurements show a P_m of 0.45. The random configuration evident in spectrum (b), however, does not reflect the stereochemistry of the precursor poly(1-fluoro-1-chloroethylene) but is rather the result of racemization at the α—carbon during the reduction. The isoregic polymer showed a greater degree of crystallinity and a higher melting point, 210°, compared to the normal polymer (190°).

The fraction of inverted units in the polymer of spectrum (a), confirmed by the ^{13}C spectrum (Tonelli *et al.*, 1981), is *ca.* 11%. When vinyl fluoride is polymerized under γ-radiation in a urea canal complex at −78°C (R. E. Cais and J. Kometani, private communication, 1982), a procedure known to encourage chain regularity, this fraction was reduced to *ca.* 5%. At the same time, a marked syndiotactic bias was introduced, P_m being about 0.35.

6.2.3.2 Poly(vinylidene fluoride)

Poly(vinylidene fluoride) has long been known but has been discovered only fairly recently to have extraordinary electrical properties. In 1969, Kawai (1969) reported its piezoelectric behavior. In 1971, Bergman *et al.* (1971) and Nakamura and Wada (1971) reported it to be pyroelectric. Both properties result from the macroscopic electric polarization of polymer samples, which transforms them into electrets. As a result of these findings, poly(vinylidene fluoride) has become one of the most intensively studied of polymers, particularly by crystallographers and physicists. Every item of knowledge, including the most accurate estimates of anomalous structures, has now become important. The occurrence of inverted monomer units was observed using ^{19}F NMR at 56.4 MHz by Wilson (1963; Wilson and Santee, 1965), who reported three resonances upfield from the head-to-tail resonance and interpreted them in detail by reference to model compounds. Liepins *et al.* (1978) also reported ^{19}F observations of inverted units. These conclusions were confirmed by ^{13}C NMR results reported by Bovey *et al.* (1977). Ferguson and Brame reported the 188 MHz ^{19}F spectrum, in which very small additional resonances not reported by Wilson could be observed. A very similar 188 MHz spectrum was observed by Tonelli *et al.* (1982) and is shown in Fig. 6.4. The principal defect resonances arise from inverted units, and correspond to the fluorines marked *A, B, C,* and *D*:

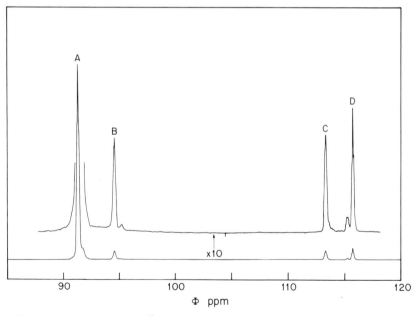

Fig. 6.4. The 188.2 MHz ^{19}F spectrum of commercial poly(vinylidene fluoride), observed in 11% (w/v) solution in dimethylformamide-d_7 at 21° (Tonelli *et al.*, 1982).

$$\overset{A}{\cdots -\underset{\longrightarrow}{CH_2-CF_2}-\underset{\longrightarrow}{CH_2-CF_2}-\overset{C}{\underset{\longleftarrow}{CF_2-CH_2}}-\overset{D}{\underset{\longrightarrow}{CH_2-CF_2}}-\overset{B}{\underset{\longrightarrow}{CH_2-CF_2}}-\cdots}$$

The fact that all four are of approximately equal intensity demonstrates that they correspond to single inversions rather than runs of reversed units. The proportion of such units varies from 3 to 6%, increasing with the temperature of polymerization. The smaller peaks (marked by vertical arrows) are believed (Tonelli *et al.*, 1982) to arise from pairs of inverted units and from head-to-head junctions, the latter presumably arising from radical recombination. These assignments cannot be regarded as established with certainty.

It has been observed by R. E. Cais and J. Kometani (private communication, 1982) that the proportion of these defects may be reduced by polymerization in urea, as with poly(vinyl fluoride), and that they may be eliminated by the following route:

$$CCl_2{=}CF_2 \xrightarrow[\text{initiator}]{\substack{\text{free}\\\text{radical}}} {-}(CCl_2{-}CF_2{)}_n \xrightarrow{Bu_3SnH} {-}(CH_2{-}CF_2{)}_n$$

The physical and electrical properties of the resulting isoregic polymer remain to be fully evaluated.

6.2.3.3 *Polytrifluoroethylene*

The structure of polytrifluoroethylene has received only limited study, but it is clear that it is irregular with respect to head-to-head versus head-to-tail addition, and with respect to stereochemical configuration as well. The ^{19}F NMR spectrum is complex and not readily interpreted in detail (Naylor and Lasoski, 1960; Wilson, 1968). R. E. Cais and J. Kometani (private communication, 1982) have reported a study analogous to those of poly(vinyl fluoride) and poly(vinylidene fluoride) which we have discussed above. They found that polytrifluorochloroethylene, in which the occurrence of inverted units may be expected to be minimal, can be made to undergo reductive dechlorination:

$$-(CF_2{-}CFCl{)}_n \xrightarrow{Bu_3SnH} -(CF_2{-}CFH{)}_n$$

The product is stereoirregular (again reflecting racemization during reduction) but nearly isoregic. In Fig. 6.5 the 188.2 MHz ^{19}F spectrum of the CF_2 groups of the "normal" (a) and isoregic (b) polymers are

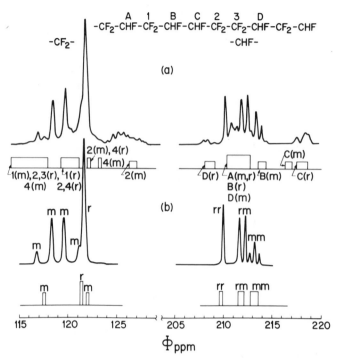

Fig. 6.5. The 188.2 MHz ^{19}F spectra of (a) "normal" and (b) isoregic polytrifluoroethylene, observed in acetone solution at 18°. The CF_2 region is on the left and the CFH region on the right (R. E. Cais and J. Kometani, private communication, 1982).

compared in the CF_2 region (left) and CFH region (right). The CF_2 resonances exhibit a classic pattern of *AB* quartet and singlet, corresponding to meso and racemic dyads:

meso   *racemic*

The geminal fluorine coupling is 289 Hz, whereas the vicinal couplings are very small and do not complicate the spectrum. The additional resonances in (a), assigned as shown in Fig. 6.5 (Tonelli *et al.*, 1982), arise from the head-to-head and tail-to-tail CF_2 and CFH groups. They are more readily evaluated quantitatively in the right-hand spectra, where the tail-to-tail CHF resonance of an inverted unit (C) appears upfield and well separated from the rest of the spectrum. It is found that the proportion of inverted units increases from 8.8% in polymer prepared at −80° to 11.6% at 0° and 12.9% at 80°, corresponding to an additional enthalpy of activation of about 200 cal-mole^{-1} for head-to-head:tail-to-tail propagation, normal propagation being favored by about 5.0 e.u. per mole. Evaluation of weak upfield resonances in Fig. 6.5b shows that polytrifluorochloroethylene itself has about 1.9% inverted units.

Copolymers of trifluoroethylene with vinylidene fluoride have ferroelectric properties similar to and even improved over those of poly(vinylidene fluoride) but those of regioregular polytrifluoroethylene are as yet unknown.

The occurrence of head-to-tail versus head-to-head:tail-to-tail (or isoregic versus syndioregic) propagation may be treated as a type of copolymerization. Defining head and tail somewhat arbitrarily, we may write for trifluoroethylene

$$\cdots-\underset{1}{\overset{t}{\text{CHF}}-\overset{h}{\text{CF}_2}}\cdot \ + \ \underset{1}{\text{CHF}=\text{CF}_2} \ \xrightarrow{k_{11}} \ \cdots-\underset{(h \text{ to } t)}{\text{CHF}-\text{CF}_2-\text{CHF}-\text{CF}_2}\cdot \qquad (6.1)$$

$$\cdots-\underset{1}{\text{CHF}-\text{CF}_2}\cdot \ + \ \underset{2}{\text{CF}_2=\text{CHF}} \ \xrightarrow{k_{12}} \ \cdots-\underset{(h \text{ to } h)}{\text{CHF}-\text{CF}_2-\text{CF}_2-\text{CHF}}\cdot \qquad (6.2)$$

$$\cdots-\underset{2}{\text{CF}_2-\text{CHF}}\cdot \ + \ \underset{1}{\text{CHF}=\text{CF}_2} \ \xrightarrow{k_{21}} \ \cdots-\underset{(t \text{ to } t)}{\text{CF}_2-\text{CHF}-\text{CHF}-\text{CF}_2}\cdot \qquad (6.3)$$

$$\cdots-\underset{2}{\text{CF}_2-\text{CHF}}\cdot \ + \ \underset{2}{\text{CF}_2=\text{CHF}} \ \xrightarrow{k_{22}} \ \cdots-\underset{(t \text{ to } h)}{\text{CF}_2-\text{CHF}-\text{CF}_2-\text{CHF}}\cdot \qquad (6.4)$$

The product of reaction 6.4 cannot be distinguished from that of 6.1 (except perhaps by examination of chain end groups) and it is probable that both $\cdots-\text{CHF}-\text{CF}_2\cdot$ and $\cdots-\text{CF}_2-\text{CHF}\cdot$ are chain carriers (Tedder et al., 1971; Sloan et al., 1975). Usually, of course, head-to-tail is strongly predominant, and this in fact is the usual basis for defining which is head and which is tail. In trifluoroethylene polymerization it appears that k_{12} is comparable to k_{11}, and that k_{21} has a similar relationship to k_{22}. If $r_1 < 1$ and $r_2 < 1$, so that $r_1 r_2 \ll 1$, the system will tend toward the alternating or crossover type, which in this case means syndioregic. It will be strictly syndioregic if $r_1 r_2 = 0$.

6.3 BRANCHING IN VINYL POLYMERS

Branching is a highly important structural variable that has received substantial theoretical and experimental study, but only very limited observation and measurement in the detailed manner we have described for stereochemical configuration and copolymer sequences. Only polyethylene and poly(vinyl chloride) have received such attention and we shall describe this work.

Branching may be produced deliberately by the introduction of dienes and divinyl or polyvinyl monomers as comonomers. These comonomers yield double bonds in the copolymer chains, which can then polymerize to yield branches and crosslinks:

1,2-Butadiene unit; 1,4-butadiene unit

p-Divinylbenzene

We speak rather here of branching introduced by processes that are under less specific control and involve chain transfer reactions of various types. Such reactions are particularly to be expected for highly reactive polymer radicals that are not stabilized by resonance, such as those from ethylene, vinyl chloride, and vinyl acetate.

6.3.1 Polyethylene

Short branches are of particular importance in polyethylene, as their presence reduces the melting point and extent of crystallinity. High pressure polyethylene has long been recognized as having short branches, but the measurement of their numbers and type has engendered some disagreement and controversy. The traditional method for their observation is by measurement of the CH_3 distortion band at 1375 cm^{-1} in the infrared spectrum in Fig. 2.9, which measures the total content of methyl groups at branch ends. This band strongly overlaps the CH$_2$ wagging bands at 1350—1360 cm^{-1}. Earlier measurements did not correct properly for this overlap and tended to overestimate the branch frequency. Willbourn (1959), showed that by using the linear material as a blank, correct quantitative results could be obtained. However, their length, type— i.e., whether trifunctional or tetrafunctional—and distribution are not revealed by vibrational spectroscopy.

We have seen in Chapter 3 that the ^{13}C chemical shifts of paraffinic hydrocarbons are highly sensitive to the proximity of tertiary carbons—that is, branch points. There is a comparable sensitivity to branch and chain ends. This sensitivity to details of local structure has been exploited to provide a fairly complete picture of the branch structure of high pressure polyethylene (Dorman *et al.*, 1972; Randall, 1973, 1978; Cudby and Bunn, 1976; Bovey *et al.*, 1976, 1979; Cheng *et al.*, 1976; Axelson *et al.*, 1977, 1979; Bowmer and O'Donnell, 1977). In Fig. 6.6 is shown the 50 MHz ^{13}C NMR spectrum of a branched polyethylene; the resonances are labeled according to the scheme inset in the figure. The assignments are made on the basis of a body of empirical chemical shift rules (Spiesecke and Schneider, 1961; Grant

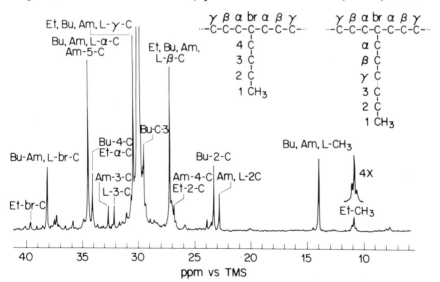

Fig. 6.6. The 50 MHz ^{13}C NMR spectrum of high-pressure polyethylene, observed in 5% (w/v) solution in 1,2,4-trichlorobenzene at 110° (F. C. Schilling, private communication, 1981).

and Paul, 1964; Grant and Cheney, 1967; D. E. Dorman, R. E. Carhart, and J. D. Roberts, private communication, 1971; Bovey, 1975) and the spectra of polyethylenes with known types of branches prepared by copolymerization with olefins (Dorman, *et al.*, 1972; Randall, 1973). The principal peak, at 30.0 ppm, not shown at its full height, corresponds to those methylene groups that are four carbons or more removed from a branch or chain end, and constitutes about 80% of the spectral intensity. The C_1 carbons (i.e., methyl groups) and C_2 carbons are the most shielded, branch point carbons the least. Main-chain carbons β to the branch are more shielded while those α to the branch are less shielded than unperturbed methylenes. The composition of this polyethylene is shown in Table 6.1. The

TABLE 6.1

Branching in High-Pressure Polyethylene

$$\overline{M}_n = 18,400, \quad \overline{M}_w = 129,000$$

Types of branch	Number of branches per 1000 backbone carbons
-CH$_3$	0.0
-CH$_2$CH$_3$	1.0
-CH$_2$CH$_2$CH$_3$	0.0
-CH$_2$CH$_2$CH$_2$CH$_3$	9.6
-CH$_2$CH$_2$CH$_2$CH$_2$CH$_3$	3.6
Hexyl and longer	5.6
Total	19.8

predominant branch type is thus *n*-butyl. Both amyl and butyl branches are believed to be formed by intramolecular chain transfer or "backbiting", as proposed by Roedel (1953):

$$
\begin{array}{ccc}
-\text{CH}_2 \quad \cdot\text{CH}_2 & \quad & -\text{CH}\cdot \quad \text{CH}_3 \\
\diagdown_{(\text{CH}_2)_n}\diagup & \longrightarrow & \diagdown_{(\text{CH}_2)_n}\diagup
\end{array}
\qquad (6.5)
$$

It may be concluded that this reaction is most probable when $n = 3$ or 4, has a low but finite probability when $n = 1$, and zero probability when $n = 0$ or 2.

The ethyl branch resonances in Fig. 6.6 require some further discussion. Ethyl branches were once thought to predominate but can now be seen to be present only to a minor extent. The methyl resonance at *ca.* 11 ppm exhibits fine structure and is not the sharp singlet observed in ethylene-*n*-butene copolymers (Dorman *et al.*, 1972; Randall, 1973; Bovey *et al.*, 1976). The resonance at 39.5 ppm, the appropriate position for the branch-point carbon of isolated ethyl branches, is much too weak in comparison to the methyl resonance. There are in addition a dozen or so small unassigned resonances in the $37-39$ ppm region. These observations suggest that there are several kinds of ethyl branches. The Roedel backbiting mechanism has been extended by Willbourn (1959) to account for the formation of ethyl branches, either as pairs of ethyl groups in a 1,3 relationship, which presumably may be meso or racemic, or as 2-ethylhexyl branches (see following page). A difficulty with this scheme is that in order to generate paired ethyl groups or 2-ethylhexyl groups at a detectable level, one must suppose that the probability of the second backbite, P_2, is considerably greater than that of the first, P_1.

The branches described in Table 6.1 as "hexyl or longer" are believed to be truly long; their frequency is estimated from the intensity of the L-3-C peak at 32.0 ppm. Such branches are believed to be formed by intermolecular chain transfer:

$$\cdot\cdot-CH_2-CH_2-CH_2-CH_2\cdot \ + \ \cdot\cdot-CH_2-CH_2-CH_2-CH_2-CH_2-\cdot\cdot$$

(6.6a)

$$\longrightarrow \ \cdot\cdot-CH_2-CH_2-CH_2-CH_3 \ + \ \cdot\cdot-CH_2-CH_2-\overset{\bullet}{C}H-CH_2-CH_2-\cdot\cdot$$

$$\cdot\cdot-CH_2-CH_2-\overset{\bullet}{C}H-CH_2-CH_2-\cdot\cdot$$

$$\xrightarrow{\text{monomer}} \ \cdot\cdot-CH_2-CH_2-\underset{\underset{\underset{\underset{\underset{CH_3}{|}}{CH_2}}{\vdots}}{\underset{|}{CH_2}}}{\overset{|}{C}H}-CH_2-CH_2-\cdot\cdot \qquad (6.6b)$$

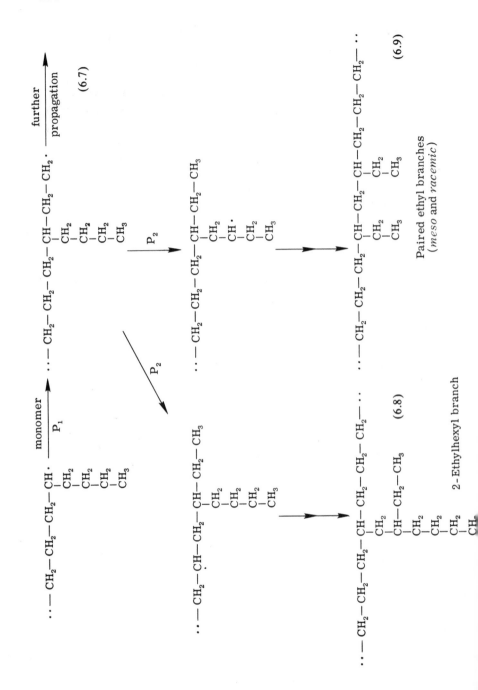

This conclusion cannot be demonstrated by NMR, since all branches of six carbons or longer are spectroscopically indistinguishable, but is supported by classical studies employing solution viscosity, gel permeation chromatography, and light scattering (Bovey et al., 1976; Axelson et al., 1977; Foster, 1979), which yielded long branch frequencies in approximate agreement with the ^{13}C results. (Note that for the polymer in Table 6.1, if we assume the presence of one methyl chain end in the unbranched chain, there are three "long" branches per number average molecule.) Mattice and Stehling (1981) have calculated the probabilities of generating branches of varying lengths by backbiting, and find a surprisingly high frequency of branches of six carbons or longer, but still not truly long. If this estimate is valid, then a substantial fraction of the branches termed "long" will not actually reduce the solution viscosity or affect the gel permeation chromatogram to the expected extent, and the figure in Table 6.1 must be regarded as an overestimate.

6.3.2 Poly(vinyl chloride)

The significance of branches in poly(vinyl chloride) is different from that of the branches in polyethylene. Poly(vinyl chloride) is only marginally crystalline at most and has a markedly lower branch content. Their importance is principally as defect structures which may initiate the chemical decomposition of the polymer. Poly(vinyl chloride) is a very valuable material, used in huge volume for moldings and electrical insulation. It is fire resistant and relatively cheap, but when heated is subject to loss of hydrogen chloride, leading to the formation of deeply colored polyene structures:

$$\cdots\!-\!CH_2CHClCH_2CHCl\!-\!\cdots \xrightarrow{\Delta} \cdots\!-\!CH\!=\!CHCH\!=\!CH\!-\!\cdots \ + \ 2\,HCl$$

The reaction is more complicated than this simple representation suggests and may be sensitized by the presence of labile centers such as tertiary chlorine. It is catalyzed by the hydrogen chloride produced, and in practice the polymer is stabilized by additives that, among other functions, absorb the HCl (Loan and Winslow, 1972; Braun, 1981; Starnes, 1981).

The direct observation of the ^{13}C spectrum of poly(vinyl chloride) is not an effective means for the observation of anomalous structures because of the multiplicity of resonances arising from stereochemical configuration (Fig. 3.27). The stereochemistry can be abolished by reductive dehalogenation with lithium aluminum hydride, $LiAlH_4$ (Cotman, 1953, 1955; George et al., 1958), or preferably the much

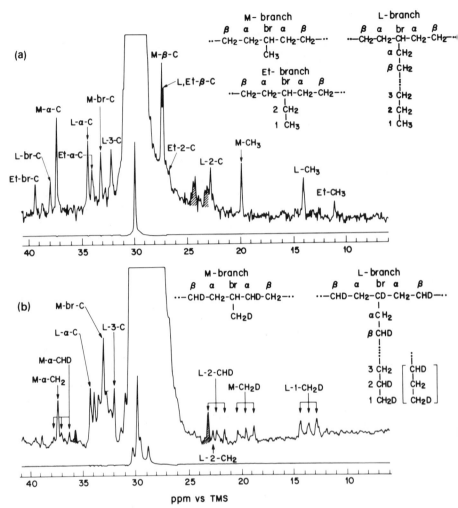

Fig. 6.7. The 25 MHz ^{13}C NMR spectra of reduced poly(vinyl chloride) observed in 5% (w/v) solution in 1,2,4-trichlorobenzene at 110°; (a) reduced with tri(*n*-butyl)tin hydride; (b) reduced with tri(*n*-butyl)tin deuteride (F. C. Schilling, private communication, 1981).

more convenient and effective free radical reagent tri-*n*-butyltin hydride, Bu₃SnH (Starnes *et al.*, 1977, 1979a). The resulting hydrocarbon chains can then be analyzed in the same manner as polyethylene (Abbås *et al.*, 1975). (The use of this reagent has already been referred to in Section 6.2.1.) One can mark the positions of the departing chlorine atoms by using the deuterium versions of these reagents, which replace the chlorine by deuterium instead of a proton. The deuterium splits the resonance of the carbon to which it is bonded to a 1:1:1 triplet (spacing *ca.* 19 Hz) and also causes an upfield shift of about 0.5 ppm (Bovey *et al.*, 1975). For small peaks the presence of deuterium essentially abolishes the carbon resonance. In Fig. 6.7a is shown the 25 MHz ^{13}C spectrum of poly(vinyl chloride) prepared at 100° and reduced with Bu_3SnH; for the reduced polymer, $\overline{M}_n = 5600$ and $\overline{M}_w = 11900$. The principal short branch (2−3 per 1000 main-chain carbons) is CH_3, with smaller proportions of ethyl, butyl (not visible in this spectrum), and long branches. In Fig. 6.7b is shown the spectrum of the same polymer reduced with Bu_3SnD. This spectrum contains more information than 6.7a. The low gain spectrum at the bottom shows the CH_2 singlet and the upfield-shifted CHD triplet. (Even the CH_2 is observably shifted upfield, by *ca.* 2-3 Hz, owing to a next-nearest neighbor isotope effect.) The upper spectrum at higher gain reveals that the methyl branches in spectrum 6.7a are actually chloromethyl groups. Spectrum 6.7b also provides much additional information, which we shall discuss in more detail a little later.

$$\cdots -CH_2-\underset{\underset{Cl}{|}}{CH}-CH_2-\underset{\underset{Cl}{|}}{CH}\cdots$$

(I)

Riga *et al.* (6.10a)   Overberger (6.11)

$CH_2=CHCl$

$$\cdots -CH_2-\underset{\underset{Cl}{|}}{CH}-CH_2-\underset{\underset{Cl}{|}}{CH}-\underset{\underset{Cl}{|}}{CH}-CH_2\cdots \xrightarrow[(6.10b)]{O} \cdots -CH_2-\underset{\underset{Cl}{|}}{CH}-\underset{\underset{Cl}{|}}{CH}-CH_2$$

(II)   (III)

$CH_2=CHCl$ (6.12)

$$\cdots -CH_2-\underset{\underset{Cl}{|}}{CH}-\overset{\overset{CH_2Cl}{|}}{CH}-CH_2-\underset{\underset{Cl}{|}}{CH}-\cdots$$

(IV)

Let us now consider the mechanism by which chloromethyl branches are formed. Rigo *et al.* (1972) proposed that they were formed by an occasional head-to-head addition followed by a 1,2

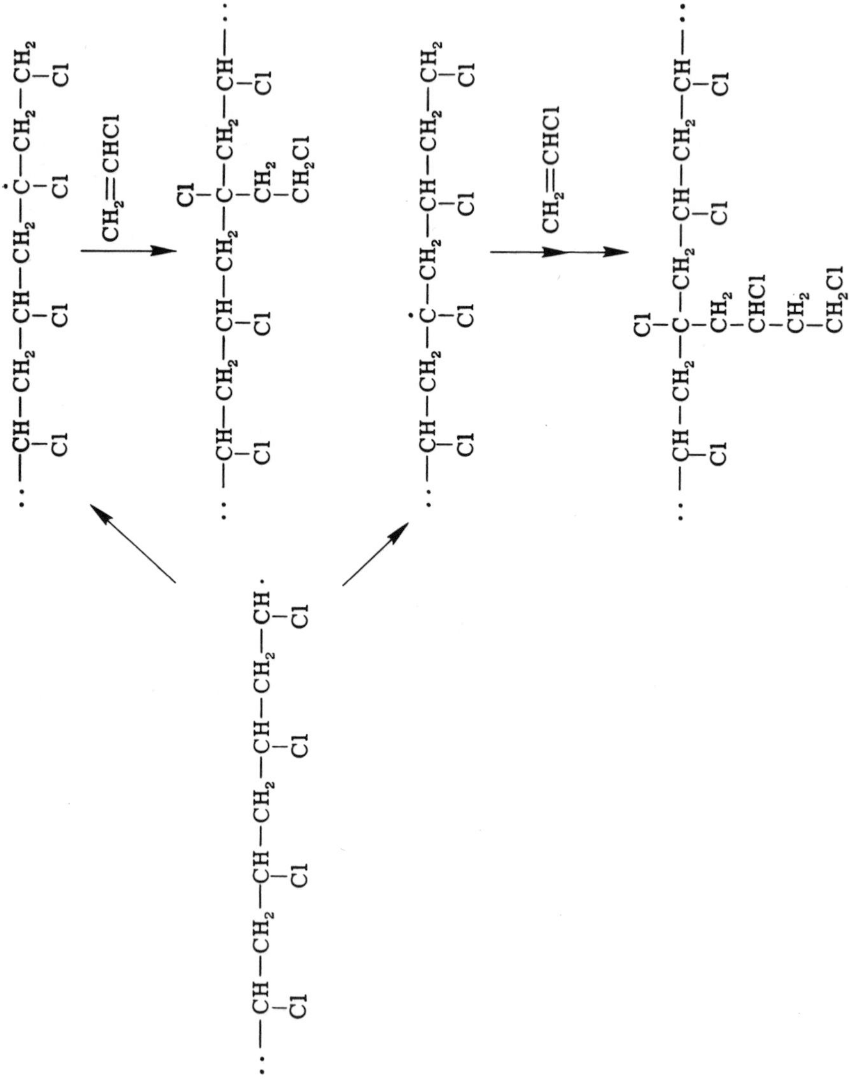

$(6.13a)$

$(6.13b)$

$(6.14a)$

$(6.14b)$

$$\cdots -CH-CH_2-CH-CH_2-\overset{\cdot}{C}-CH_2-CH_2-Cl$$
$$\quad\; |Cl \quad\; |Cl \quad\; |Cl$$

$$\xrightarrow{CH_2=CHCl}$$

$$\cdots -CH-CH_2-CH-CH_2-C-CH_2-CH-\cdots$$
$$\quad\; |Cl \quad\; |Cl \quad\; |CH_2 \quad |Cl$$
$$\qquad\qquad\qquad\qquad\quad |CH_2Cl$$

$$\cdots -CH-CH_2-CH-CH_2-CH-CH_2-CH_2-Cl$$
$$\quad\; |Cl \quad\; |Cl \quad\; |Cl$$

$$\xrightarrow{CH_2=CHCl}$$

$$\cdots -CH-CH_2-CH-CH_2-CH-CH_2-CH-\cdots$$
$$\quad\; |Cl \quad\; |Cl \quad\; |CH_2 \quad\; |Cl$$
$$\qquad\qquad\qquad\qquad\quad |CHCl$$
$$\qquad\qquad\qquad\qquad\quad |CH_2$$
$$\qquad\qquad\qquad\qquad\quad |CH_2Cl$$

$$\cdots -CH-CH_2-CH-CH_2-CH-CH_2-CH\cdot$$
$$\quad\; |Cl \quad\; |Cl \quad\; |Cl \quad\; |Cl$$

chlorine shift to form the more stable secondary radical (III). C. G. Overberger (private communication, 1974) suggested that radical II was formed directly by a hydrogen shift in the normal propagating radical (I). Both mechanisms are consistent with the observed final structure IV, and both pass through the same radical III. However, according to the Rigo mechanism the β−carbon of the monomer becomes the methylene group of the chloromethyl branch, whereas according to the Overberger mechanism the α−carbon performs this role. By employing α−deuterovinyl chloride, it has been clearly shown that the former route is the correct one (Starnes et al., 1979b). This finding incidentally disposes of the question of head-to-head addition in vinyl chloride propagation; head-to-head units have not been directly detected.

1-Chloroethyl and 1,3-dichlorobutyl branches are presumed to be formed by intramolecular backbiting chain transfer, in which the α−protons are removed (see facing page). The occurrence of 1,3-dichlorobutyl branches at a level of 0.6−1.0 per 1000 backbone carbons can be detected by careful examination of reduced poly(vinyl chloride) prepared under normal bulk polymerization conditions (Starnes et al., 1981a). They can be much more readily observed in poly(vinyl chloride) prepared under "starved" conditions, e.g., when polymerization occurs in a solvent and the growing radical has more time to backbite, in competition with the addition of monomer.

Long branches are very probably formed by intermolecular chain transfer reactions parallel to Eqs. (6.13) and (6.14). It will be noted that all branches except chloromethyl have a tertiary chlorine atom at the branch point carbon and furnish labile centers at which dehydrochlorination may be initiated.

$$\cdots-CH_2-\underset{\underset{Cl}{|}}{CH}-CH_2-\underset{\underset{Cl}{|}}{CH}\cdot \; + \; \cdots-CH_2-\underset{\underset{Cl}{|}}{CH}-CH_2-\underset{\underset{Cl}{|}}{CH}-CH_2-\underset{\underset{Cl}{|}}{CH}-CH_2-\cdots \qquad (6.15a)$$

$$\cdots-CH_2-\underset{\underset{Cl}{|}}{CH}-CH_2-CH_2 \; + \; \cdots-CH_2-\underset{\underset{Cl}{|}}{CH}-CH_2-\overset{\displaystyle \cdot}{\underset{\underset{Cl}{|}}{C}}-CH_2-\underset{\underset{Cl}{|}}{CH}-CH_2-\cdots \qquad (6.15b)$$

$$CH_2{=}CHCl$$

$$\cdots -CH_2-CH-CH_2-\overset{\overset{\displaystyle Cl}{|}}{\underset{\underset{\underset{\underset{\underset{\underset{\vdots}{CHCl}}{|}}{\underset{CH_2}{|}}}{\underset{CHCl}{|}}}{\underset{CH_2}{|}}}{C}}-CH_2-\overset{}{\underset{\underset{Cl}{|}}{CH}}-CH_2-\cdots$$

In addition to these features, it is also possible to detect 1,3-dichloro and 1,2,4-trichloro chain ends, $-CH_2CHClCH_2CH_2Cl$ and $-CHClCH_2CHClCH_2Cl$, respectively. The most plausible mechanism to account for these is as follows; here, radical III (p. 179), in addition to the continued propagation represented by Eq. (6.12), also undergoes in part a process of β−scission with production of a chlorine atom:

$$\cdots -CH_2-\underset{\underset{Cl}{|}}{CH}-CH-\underset{\underset{Cl}{|}}{CH_2} \xrightarrow{\overset{}{(\longleftarrow)}} \cdots -CH_2-CH=CH-\underset{\underset{Cl}{|}}{CH_2} + Cl\cdot \quad (6.16)$$

$$(\text{III}) \qquad\qquad\qquad\qquad\qquad\qquad (\text{IV})$$

The chlorine atom then initiates propagation:

$$Cl\cdot + CH_2=CHCl \longrightarrow ClCH_2-\overset{\cdot}{C}HCl \longrightarrow \qquad (6.17)$$

$$\cdots -CHCl-CH_2-CHCl-CH_2Cl$$

$$1,2,4\text{-Trichloro } (75\%)$$

The 1,3-dichloro chain ends are assumed to be generated by chain transfer to solvent (if any), or to polymer [Eq. (6.15a)]; they constitute the other 25% of the observable chlorine substituted chain ends. [The chloroallyl end-groups generated in reaction (6.16) are not observed directly because they are converted into 1-ethyl-2-alkyl-cyclopentyl groups in the tributyltin hydride reduction (Starnes et al., 1981b).]

6.3.3 Poly(vinyl acetate)

Long branches in poly(vinyl acetate) are believed to be formed by chain transfer to the acetoxy methyl group. Such branching may be measured by hydrolysis of the ester links followed by molecular weight measurement, but this method is correct only if the mechanism is correct. Stein (1964) has made careful molecular weight measurements indicating a rapid increase in the latter stages of monomer conversion. Spectroscopic evidence would be desirable but has not yet been provided.

main chain

$$\cdots -CH_2CH-\cdots$$
$$|$$
$$O$$
$$|$$
$$C{=}O$$
$$|$$
$$CH_2-CH_2-CH-CH_2-CH-\cdots$$
$$|\qquad\quad|$$
$$O\qquad\quad O$$
$$|\qquad\quad|$$
$$C{=}O\qquad C{=}O$$
$$|\qquad\quad|$$
$$CH_3\qquad CH_3$$

} branch

REFERENCES

Abbås, K. B., Bovey, F. A., and Schilling, F. C. (1975). *Makromol. Chem., Suppl.* **1**, 227.

Alfrey, T., Jr., Hass, H. C., and Lewis, C. W. (1951). *J. Am. Chem. Soc.* **73**, 2851.

Alfrey, T., Jr., Hass, H. C., and Lewis, C. W. (1952a). *J. Am. Chem. Soc.* **74**, 2025.

Alfrey, T., Jr., Hass, H. C., and Lewis, C. W. (1952b). *J. Am. Chem. Soc.* **74**, 2097.

Axelson, D. E., Mandelkern, L., and Levy, G. C. (1977). *Macromolecules* **10**, 557.

Axelson, D. E., Levy, G. C., and Mandelkern, L. (1979). *Macromolecules* **12**, 41.

Bergman, J. G., Jr., McFee, J. H., and Crane, G. R. (1971). *Appl. Phys. Lett.* **18**, 203.

Bovey, F. A. (1975). *Proc. Int. Sym. Macromolecules, Rio de Janeiro, July 26-31, 1974* (E. B. Mano, ed.), p. 169 *et seq.* Elsevier, Amsterdam.

Bovey, F. A., Abbås, K. B., Schilling, F. C., and Starnes, W. H., Jr. (1975). *Macromolecules* **8**, 437.

Bovey, F. A., Schilling, F. C., McCrackin, F. L., and Wagner, H. L. (1976). *Macromolecules* **9**, 76.

Bovey, F. A., Schilling, F. C., Kwei, T. K., and Frisch, H. L. (1977). *Macromolecules* **10**, 559.

Bovey, F. A., Schilling, F. C., and Starnes, W. H., Jr. (1979). *Polymer Preprints* **20**(2), 160.

Bowmer, T. N., and O'Donnell, J. H. (1977). *Polymer* **18**, 1032.

Braun, D. (1981). In "Developments in Polymer Degradation-3" (N. Grassie, ed.), p. 101. Applied Science Publishers, London.

Cais, R. E., and Spencer, C. P. (1982). *Eur. Polymer J.* **18**, 189.

Cheng, H. N., Schilling, F. C., and Bovey, F. A. (1976). *Macromolecules* **9**, 363.

Cotman, J. D., Jr. (1953). *Ann. New York Acad. Sci.* **57**, 417.

Cotman, J. D., Jr. (1955). *J. Am. Chem. Soc.* **77**, 2790.

Cudby, M. E. A., and Bunn, A. (1976). *Polymer* **17**, 345.

Dorman, D. E., Otocka, E. P., and Bovey, F. A. (1972). *Macromolecules* **5**, 574.

Flory, P. J. (1939). *J. Am. Chem. Soc.* **61**, 1518.

Flory, P. J., and Leutner, F. S. (1948). *J. Polymer Sci.* **3**, 880.

Flory, P. J., and Leutner, F. S. (1950). *J. Polymer Sci.* **5**, 267.

Foster, G. N. (1979). *Polymer Preprints* **20**(2), 463.

George, M. H., Grisenthwaite, R. J., and Hunter, R. F. (1958). *Chem. Ind. (London)* **1958**, 1114.

Grant, D. M., and Cheney, B. V. (1967). *J. Am. Chem. Soc.* **89**, 5315, 5319.

Grant, D. M. and Paul, E. G. (1964). *J. Am. Chem. Soc.* **86**, 2984.

Kawai, N. (1969). *Japan J. Appl. Phys.* **8**, 1975.

Liepins, R., Surles, J. R., Morosoff, N., Stannett, V. T., Timmons, M. L., and Wortman, J. J. (1978). *J. Polymer Sci., Polymer Chem. Ed.* **16**, 3039.

Loan, L. D., and Winslow, F. H. (1972). *In* "Polymer Stabilization" (W. L. Hawkins, ed.) Chapter 3. Wiley (Interscience), New York.

Marvel, C. S., Sample, J. H., and Roy, M. F. (1939). *J. Am. Chem. Soc.* **61**, 3241.

Mattice, W. L. and Stehling, F. C. (1981). *Macromolecules* **14**, 1479.

Nakamura, K., and Wada, Y. (1971). *J. Polymer Sci. A-2* **9**, 161.

Naylor, R. E., Jr., and Lasoski, S. W., Jr. (1960). *J. Polymer Sci.* **44**, 1.

Randall, J. C. (1973). *J. Polymer Sci., Polymer Phys. Ed.* **11**, 275.

Randall, J. C. (1978). *J. Appl. Polymer Sci.* **22**, 585.

Rigo, A., Palma, G., and Talamini, G. (1972). *Makromol. Chem.* **153**, 219.

Roedel, M. J. (1953). *J. Am. Chem. Soc.* **75** 6110.

Sloan, J. P., Tedder, J. M., and Walton, J. C. (1975). *J. Chem. Soc. Perkin* **11**, 1846.

Smets, G. (1966). *Pure Appl. Chem.* **12**, 218.

Spiesecke, H., and Schneider, W. G. (1961). *J. Chem. Phys.* **35**, 722.

Starnes, W. H., Jr. (1981). *In* "Developments in Polymer Degradation-3" (N. Grassie, ed.), p. 135. Applied Science Publishers, London.

Starnes, W. H., Jr., Hartless, R. L., Schilling, F. C., and Bovey, F. A. (1977). *Polymer Preprints* **18**(1), 499.

Starnes, W. H., Jr., Schilling, F. C., Abbås, K. B., Plitz, I. M., Hartless, R. L., and Bovey, F. A. (1979a). *Macromolecules* **12**, 13.

Starnes, W. H., Jr., Schilling, F. C., Abbås, K. B., Cais, R. E., and Bovey, F. A. (1979b). *Macromolecules* **12**, 556.

Starnes, W. H., Jr., Schilling, F. C., Plitz, I. M., Cais, R. E., and Bovey, F. A. (1981a). *Polymer Bull.* **4**, 555.

Starnes, W. H., Jr., Villacorta, G. M., and Schilling, F. C. (1981b). *Polymer Preprints* **22**(2), 307.

Stein, D. J. (1964). *Makromol. Chem.* **76**, 170.

Tedder, J. M., Walton, J. C., and Winton, K. D. R. (1971). *Chem. Commun.*, 1046.

Tepelekian, M., Tho, P. Q., and Guyot, A. (1969). *Eur. Polymer J.* **5**, 795.

Tonelli, A. E., Schilling, F. C., and Cais, R. E. (1981). *Macromolecules* **14**, 560.

Tonelli, A. E., Schilling, F. C., and Cais, R. E. (1982). *Macromolecules* **15**, in press.

Vogel, O. (1979). *Polymer Preprints* **20**(1), 154.

Willbourn, A. H. (1959). *J. Polymer Sci.* **34**, 569.

Wilson, C. W., III (1963). *J. Polymer Sci., Part A* **1**, 1305.

Wilson, C. W., III (1968). Paper presented at the 155th American Chemical Society Meeting, Division of Fluorine Chemistry, San Francisco.

Wilson, C. W., III, and Santee, E. R., Jr. (1965). *J. Polymer Sci, Part C* **8**, 97.

Chapter 7

CHAIN CONFORMATIONS
OF MACROMOLECULES

7.1 INTRODUCTION

A wide variety of shapes can be assumed by flexible chains. In a long chain of connected segments, the spatial relationship between neighboring segments is governed by *bond lengths*, by *bond angle*, and by the *rotational state* of these bonds—those that can be rotated. Such rotation is possible in most macromolecular chains and leads to an astronomically large number of permissible spatial arrangements of the segments, so many that a statistical approach is absolutely necessary. We must deal with *average* quantities, such as the *average* spatial extent of the molecule (even assuming a fixed contour length) and *distribution functions*—for example, the probability that a given segment is located at a specified coordinate relative to another.

A useful quantity for measuring chain conformation is the end-to-end distance. Consider a chain of n segments, shown in Fig. 7.1.

185

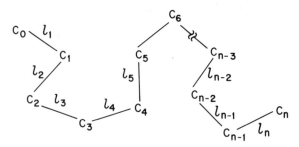

Fig. 7.1.   Schematic representation of a polymer chain backbone.

Each segment is represented by a vector l_i numbered from 1 to n. The end-to-end distance for a given conformation is simply the vector connecting the two ends:

$$\mathbf{r} = \sum_{i=1}^{n} \mathbf{l}_i \tag{7.1}$$

Ordinarily, only the scalar magnitude of $\mathbf{r}$ can be related conveniently to experimentally measured quantities. The square of the scalar quantity is

$$r^2 = \mathbf{r} \cdot \mathbf{r} = \sum_{i,j} \mathbf{l}_i \cdot \mathbf{l}_j \tag{7.2}$$

where the summation is carried out from 1 to n for i and j.

The summation in Eq. (7.2) can be written in an alternative form,

$$r^2 = \sum_i l_i^2 + 2 \sum_{i \neq j} \mathbf{l}_i \cdot \mathbf{l}_j \tag{7.3}$$

The first term is simply the product of $\mathbf{l}_i$ with itself and the second term is the scalar product of $\mathbf{l}_i$ with $\mathbf{l}_j$ when i is not equal to j. The factor of 2 results from the equivalence of $\mathbf{l}_i \cdot \mathbf{l}_j$ and $\mathbf{l}_j \cdot \mathbf{l}_i$, requiring double counting.

Another parameter that is useful in the description of the size of the macromolecule is the radius of gyration s. Consider the chain as an assembly of mass elements, each located at distance s_i from the center of gravity. The radius of gyration is defined as

$$s^2 = \frac{\sum_i m s_i^2}{\sum m} = \frac{\sum_{i=1}^{n} s_i^2}{n} \tag{7.4}$$

To obtain a statistical mechanical average of the above quantities, the averaging process is performed for all possible conformations. The average quantity is denoted by angle brackets, for example, $<r^2>$. The reference state for such a calculation is the one in which the molecule is free of external constraints, such as applied force or interaction with solvent. The *unperturbed state dimension* is denoted by the subscript 0 appended to the angle bracket, e.g., $<r^2>_0$ or $<s^2>_0$. For very long unperturbed chains it can be shown that

$$<s^2>_0 = \frac{1}{6} <r^2>_0 \tag{7.5}$$

7.2 THE FREELY JOINTED CHAIN

The *freely jointed chain* is a hypothetical linear chain where all the segments are of equal length and the angle between two successive segments can assume any value. Bond rotation is also free. For such a hypothetical chain, where there is no correlation between the direction of neighboring bonds and the angle θ between two successive bonds can assume any value with equal probability, the average of $\mathbf{l}_i \cdot \mathbf{l}_j$, i.e., the average of $l_i l_j \cos \theta$ over all possible values of θ, is zero when $i \neq j$. Summing over i and j values from 1 to n in Eq. (7.3), we obtain

$$<r^2>_0 = nl^2 \tag{7.6}$$

Although a real polymer chain is far from freely jointed, the correlation between bonds i and $i + j$ must vanish as j increases. Consequently, it is sometimes useful to represent the real chain as an ensemble of "statistical segments" forming an equivalent freely jointed chain, subject to the condition that the mean square end-to-end distance and the fully extended length r_{max} of this hypothetical chain equal those of the real chain,

$$<r^2>_0 = n'l'^2$$

and

$$r_{max} = n'l'$$

Regardless of the exact nature of the conformational restrictions in a real chain, the proportionality between the unperturbed dimensions, as measured by $<r^2>_0$, and nl^2 is always maintained. The *characteristic ratio* $<r^2>_0/nl^2$, sometimes designated as C_∞ (see Section 7.4), is greater than unity for all real chains, and is a measure of the departure from a freely jointed or freely rotating model. The theoretical calculation of C_∞ is discussed in Section 7.4 and its experimental measurement in Section 7.5.

7.3 THE FREELY ROTATING CHAIN

If the angles between successive bonds in a chain are held fixed but the rotation of the bond is free, the resulting chain is termed *freely rotating*. Unlike the case of the freely jointed chain, the average of the scalar product of l_i with l_j is no longer zero. The projection of bond $i + 1$ on bond i is $l \cos \theta$ and

$$<l_i \cdot l_{i+j}> = l^2 (\cos \theta)^j \tag{7.7}$$

(Note that the traverse projection of l_{i+1} on l_i averages to zero because of free rotation.) Substitution into Eq. (7.3) yields, for large n,

$$<r^2>_0 \cong \left[\frac{1-\cos \theta}{1+\cos \theta} \right] nl^2 \tag{7.8}$$

$$= 2nl^2 \text{ for a tetrahedral chain} \tag{7.9}$$

7.4 CHAINS WITH RESTRICTED ROTATION

For real polymer chains there are intrachain hindrances that restrict each bond to a small number of distinguishable rotational states. If the relative energies of these states are known, the conformational partition function can be computed by matrix methods and from it any conformation-dependent property of the chain can be calculated. These include the mean square end-to-end distance, the dipole moment, and a number of optical properties [see General References: Vol'kenshtein (1963), Birshtein and Ptitsyn (1966), Flory (1969), Tonelli (1977)]. We now wish to examine these conformations and their energies in greater detail with respect to both computation and experimental observation.

Spectroscopic and electron diffraction studies of a variety of small molecules have demonstrated the nature of the hindering potentials that one may expect to be present also in macromolecular chains. It is well known that rotation about the C—C bond in ethane is characterized by a symmetric threefold potential with energy minima corresponding to staggering of the C—H bonds, the depths of the wells being ~3 kcal. All the staggered conformers are equivalent. In n-butane, three staggered conformers are present about the central (C_2—C_3) bond, corresponding also to a threefold potential (Fig. 7.2), which in this case, however, is not symmetric, the gauche conformers g^+ and g^- having about 0.5 kcal higher energy than the trans t:

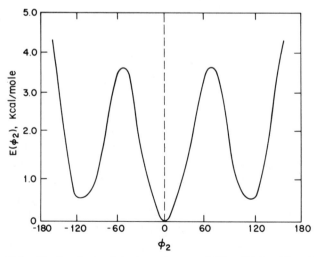

The g^+ and g^- conformers are mirror images, identical in energy, and are formally generated by rotating about the C_2—C_3 bond through an angle ϕ of $+$ and $-$ 120°, i.e., clockwise and counterclockwise, respectively, as indicated above (looking from C_3 toward C_2). The depths of the wells are such that, although small oscillations of ϕ of $\pm$ ~15—20° occur, the populations of the eclipsed conformers corresponding to the tops of the barriers are negligible; yet the barriers are low enough so that the conformers interchange with each other very rapidly, their lifetimes being of the order of 10^{-10} sec at ordinary temperatures.

Fig. 7.2. Torsional potentials about the central (C_2—C_3) bond in n-butane.

When we lengthen the chain by one carbon further complications develop. The conformations must now be specified by two rotational angles ϕ_1 and ϕ_2; here we look from C_3 toward C_4 (top) and then from C_3 toward C_2 and C_4 (bottom):

$$tt$$
$$\phi_1 = \phi_2 = 0°$$

$$tg^+$$
$$\phi_1 = 0°;$$
$$\phi_2 = +120°$$

$$tg^-$$
$$\phi_1 = 0°;$$
$$\phi_2 = -120°$$

$$g^+g^+$$
$$\phi_1 = \phi_2 = +120°$$

$$g^+g^-$$
$$\phi_1 = +120°$$
$$\phi_2 = -120°$$

We now encounter a most important principle in calculating the conformations of polymer chains: the conformational preference at a given bond is influenced by the conformations of the neighboring bonds. Thus, the energy of a given rotational state of the C_2-C_3 bond depends on the rotational state of the C_3-C_4 bond and vice versa. Relative to the *tt* state, the energies of the mirror image tg^+ and tg^- states (and of the equivalent g^+t and g^-t states) are only slightly higher than for the g^+ and g^- states of butane. The g^+g^+ ($= g^-g^-$) and g^+g^- ($= g^-g^+$) states are very different in energy; in the g^+g^+ state, the methyl groups experience an approximately neutral (i.e., neither repulsive nor attractive) interaction, while in the g^+g^- state the methyls experience a severe repulsive interaction that strongly tends to exclude this state. Calculations (Abe *et al.*, 1966) of pentane conformational energies based on semiempirical interatomic potentials, both attractive and repulsive, are represented as contours in Fig. 7.3. We can see that the exact-staggered $g^{\pm}g^{\mp}$ ($\phi_1, \phi_2 = \pm 120°, \pm 120°$) conformations are of very high energy indeed, but that there are local minima near this conformation at $\phi_1, \phi_2 = 77°, -115°$ and $120°, -70°$. The right side of the surface corresponds to $g^{\pm}g^{\pm}$ conformers. The adjusted $g^{\pm}g^{\mp}$ conformers are

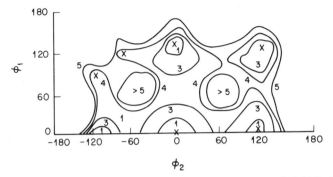

Fig. 7.3. Energy map for *n*-pentane calculated by Abe *et al.* (1966) for internal rotations about bonds 1 and 2. Methyl groups are in minimum energy staggered conformations. Contours are at intervals of 1 kcal mole^{-1}, minima being indicated by X.

about 3200 cal above *tt* and 2000 cal above $g^\pm g^\pm$. Therefore $g^\pm g^\pm$ conformers are about 1000 cal above *tt*, corresponding to two gauche interactions without pentane interference effects.

Clearly, both three-bond or *first-order* interactions, like those in *n*-butane, and the additional four-bond or *second-order* interactions encountered in *n*-pentane and longer chains, must be taken account of even in the simplest of polymer chains such as those of strictly linear polyethylene (often termed "polymethylene"). We have seen that, to a good degree of approximation, only the states of nearest-neighbor bonds need be considered, higher-order interactions being ordinarily negligible. Such a chain corresponds to the mathematical concept termed a one-dimensional Ising system.

If all bond lengths and valence angles are fixed, the conformation of a chain of *n* bonds can be specified by assigning a rotational state to each of the $n-2$ nonterminal bonds. This mode of treatment of chain statistics is called the *rotational isomeric state* (RIS) model, and clearly has the advantage of being much more realistic than the idealized models treated in Sections 7.2 and 7.3. If there are ν rotational states about each bond, there will be ν^{n-2} possible conformational states. For the common case that $\nu=3$, even a chain of only 20 bonds will have over one billion conformational states; chains of polymeric length would seem to present insuperable computational obstacles. However, they can be readily treated by matrix methods, employing a digital computer.

Consider the segment of a *polymethylene* chain shown below:

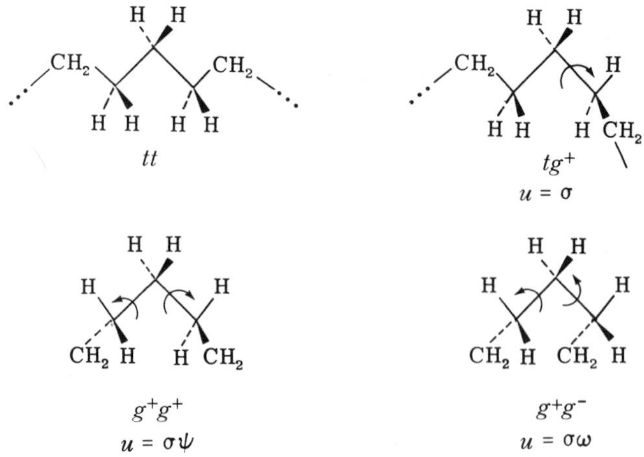

We define *statistical weights* for these conformations as Boltzmann factors

$$u_i = \exp[-E_i(\phi_i-1,\phi_i)/RT]$$

where E_i is the energy of the conformer corresponding to the rotational angles ϕ_{i-1} and ϕ_i for neighboring bonds. It is convenient to represent the statistical weights of specific conformations by symbols: tt is taken as having a statistical weight of 1; the tg^+ ($= g^+t = tg^- = g^-t$ for a polymethylene chain) conformation is assigned a statistical weight σ, the $g^{\pm\pm}$ conformation a statistical weight $\sigma\psi$, and the $g^{\pm}g^{\mp}$ conformation (almost excluded) a statistical weight $\sigma\omega$. The last two are compounded of two symbols because the conformations involve both the simple "butanelike" gauche energy and a new interaction. From what we have said before, however, concerning the methyl group interactions in the $g^{\pm}g^{\pm}$ conformation of pentane, i.e., that they are neither attractive nor repulsive, we may assume that $\psi \simeq 1$. The statistical weight matrix for two successive bonds, including both first- and second-order steric interactions, is then

$$U_i = i - 1 \begin{array}{c} \\ (t) \\ (g^+) \\ (g^-) \end{array} \overbrace{\begin{bmatrix} \widehat{(t)} & (g^+) & (g^-) \\ 1 & \sigma & \sigma \\ 1 & \sigma & \sigma\omega \\ 1 & \sigma\omega & \sigma \end{bmatrix}}^{i} \qquad (7.10)$$

Here, the columns are indexed on the bond i and the rows on the preceding bond. Since $E_\sigma - E_t$ (from butane) $\simeq 500$ cal mole^{-1}, σ (at $300°K$) $\simeq 0.43$; $E_\omega \cong 3200 - 2E_\sigma \cong 2200$ cal mole^{-1}, and therefore $\omega \simeq 0.026$.

The statistical weight of a particular conformation of a chain of n bonds is

$$\Omega_{(\phi)} = \prod_{i=2}^{n-1} u_i \qquad (7.11)$$

the first and last bonds being excluded from consideration since they are without effect. The conformational ("configurational") partition function is obtained by summing over all possible conformations:

$$Z = \sum_{(\phi)} \prod_{i=2}^{n-1} u_i \qquad (7.12)$$

Application of matrix methods (Flory, 1969) leads to

$$Z = J^* \left[\prod_{i=2}^{n-1} U_i \right] J \qquad (7.13)$$

where J^* and J are the row and column vectors:

$$J^* = [1,\sigma,\sigma] \; ; J = \begin{bmatrix} 1 \\ 1 \\ 1 \end{bmatrix} \qquad (7.14)$$

Equations (7.13) and (7.14) allow one to calculate the conformational partition function provided the bond rotational states and their energies can be estimated.

By further matrix operations, which we shall not detail here, any conformation-dependent property of a polymer chain can be calculated.

TABLE 7.1

Calculated and Experimental Values of the Characteristic Ratio
$C_\infty = <r^2>_0/nl^2$ (300°K) and its Temperature Coefficient
for the Polymethylene Chain

Model	Characteristic ratio C_∞	Temperature coefficient $dC_\infty/dT \times 10^3$
Experimental	6.7	-1.1 ± 0.1 (390–420° K)
(a) Freely rotating	2.1	0
(b) 3-state RIS, $E_\sigma = 500$ cal $E_\omega = 0$ (independent rotations)	3.4	—
(c) 3-state RIS, $E_\sigma = 500$ cal $E_\omega = 2200$ cal	6.7	-1.0
(d) 3-state RIS, $E_\sigma = 800$ cal $E_\omega = 2200$ cal	8.3	-1.5
(e) 3-state RIS, $E_\sigma = 500$ cal $E_\omega = 3500$ cal	7.1	-0.7

Table 7.1 shows the experimental characteristic ratio C_∞ (Section 7.2) of the polymethylene chain calculated on various assumptions, including (a) the freely rotating chain; (b) a chain with a reasonable value for the energy of the gauche state but with the assumption of independence of the rotational states of neighboring bonds; and (c) the neighbor-dependent three-state model with the values of E_σ and E_ω indicated above, as well as higher values (d) and (e). The energies in (c), deduced from small paraffins, give the correct value of C_∞, as well as its temperature coefficient; larger values increase C_∞ beyond its experimental value but decrease its temperature coefficient.

For the *polyoxymethylene* chain, which has a two-bond repeat unit, two statistical weight matrices are involved, one for the bonds to the O atom and one for the bonds to the CH_2 group:

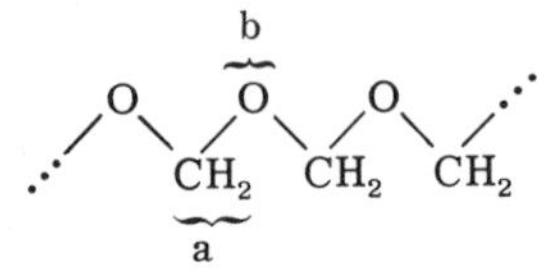

$$U_a = \begin{bmatrix} 1 & \sigma & \sigma \\ 1 & \sigma & \sigma\omega_a \\ 1 & \sigma\omega_a & \sigma \end{bmatrix} \qquad (7.15)$$

$$U_b = \begin{bmatrix} 1 & \sigma & \sigma \\ 1 & \sigma & \sigma\omega_b \\ 1 & \sigma\omega_b & \sigma \end{bmatrix} \qquad (7.16)$$

The statistical weights refer to conformations analogous to those for the polymethylene chain. The gauche conformation is known from a variety of evidence, including the known crystal structures (Section 7.6), to be of considerably lower energy than the trans, in contrast to the behavior of the polymethylene chain, presumably because of favorable coulombic interactions. There are two $g^{\pm}g^{\mp}$ conformations, corresponding to (a) and (b) below:

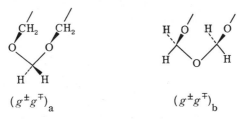

The first is very rigorously excluded by severe $CH_2 \cdots CH_2$ overlap and ω_a may be taken as 0; the corresponding $O \cdots O$ interaction in (b) is more weakly excluded and ω_b is assigned a statistical weight of ~0.05. Assuming tetrahedral geometry, a gauche angle $\phi \simeq 120°$, and a value of ~12 for a σ (i.e., $E_\sigma - E_t \simeq 1500$ cal-mole^{-1}) one obtains a value of about 10 for C_∞, in agreement with the rather approximate experimental values for this polymer, and with the dipole moments for smaller model molecules. If the $g^{\pm}g^{\mp}$ conformers are allowed to contribute appreciably, only a positive value of E_σ, clearly excluded by other evidence, predicts the observed value of C_∞; a negative value would yield a very much smaller predicted C_∞.

Rotational isomeric state model calculations of *polyoxyethylene* chains (Mark and Flory, 1965, 1966), which have a three-atom repeat unit, require three statistical weight matrices, corresponding to c—a, a—b, and b—c bond pairs in the sense indicated:

$$\cdots \underset{a}{\diagdown} O \underset{b}{\diagup} \overset{CH_2}{\underset{CH_2}{\diagdown}} \underset{a}{\diagup} O \underset{b}{\diagdown} \overset{c}{\diagup} \overset{CH_2}{\underset{CH_2}{\diagdown}} \underset{a}{\diagup} O \underset{b}{\diagdown} \cdots$$

There are two gauche conformations with the estimated energies shown (referred to the trans conformation):

$E_\sigma = 900$ cal
($\sigma = 0.26 \pm 0.03$)

$E_{\sigma'} = -250$
($\sigma' = 1.50 \pm 0.20$)

The second is favored over trans, although less so than the gauche conformation in polyoxymethylene, whereas the other is more unfavorable than the gauche polymethylene conformation because C—O bond lengths (0.143 nm) are shorter by 0.010 nm than the C—C bond lengths in the latter. Of the two $g^{\pm}g^{\mp}$ conformations, that involving four-bond $-CH_2 \cdots CH_2-$ interactions is given a statistical weight of zero, as in polyoxymethylene and polymethylene, while that involving $-O \cdots CH_2-$ interactions is assigned an energy E_ω of 800 (± 400) cal ($\omega = 0.30 \pm 0.20$). In Table 7.2 is shown a compilation of experimental properties compared to those predicted using these

TABLE 7.2

Calculated and Experimental Conformation-Dependent Properties
for Polyoxyethylene[a]

Property	Temperature (°C)	Calculated	Observed
C_∞	40	4.8	4.8
$\left[\dfrac{dlnC_\infty}{dT}\right] \times 10^3$	60	0.18	0.23
$\left[\dfrac{<\mu^2>}{n\mu^2}\right]^b_{n\to\infty}$	25	0.58	0.58
$\left[\dfrac{dln<\mu^2>}{dT}\right]^b_{n\to\infty} \times 10^3$	25	2.5	2.6

[a] See Tonelli (1977).
[b] μ = dipole moment of C-O bond.

parameters. A number of other conformation-dependent properties, including optical properties, are also correctly predicted, and so the unperturbed conformation of this polymer appears to be well understood on the basis of a three-state rotational isomeric model.

The treatment of the conformations of vinyl homopolymer chains having asymmetric centers involves further complications in that the steric requirements of the side chains must be taken into account. There is also the formal complication that, because of the asymmetric centers, one must to some degree view the chain stereochemistry in an arbitrary manner and must generate chain conformations according to a consistent convention for bond rotations. For first-order interactions (adopting an arbitrary l chirality for the bonds about C_α), we have the following rotational states (here, we must imagine that we are looking from the $\beta-$carbon toward the $\alpha-$carbon):

$$t, \; u_1 = \sigma\eta \qquad\qquad g^-, \; u_1 = \sigma \qquad\qquad g^+, \; u_1 = \sigma\tau$$

The $C_\alpha H \cdots R$ interaction in the trans conformation is given the statistical weight $\sigma\eta$; if R is equivalent to CH_2, η will be unity and the interaction is essentially equivalent to the gauche interaction in a polymethylene chain. In general, this will not be true and η will be the ratio of the statistical weight of the first-order interaction involving R to the corresponding interaction involving CH_2. The g^- conformations involve a $CH_2 \cdots CH$ interaction and are weighted by σ. The g^+ conformations involve both $CH_2 \cdots CH$ and $CH \cdots R$ interactions and are assigned the statistical weight $\sigma\tau$. The relative statistical weights for t, g^+, and g^- are thus η, τ and 1, respectively, and may be expressed in terms of the diagonal matrix:

$$\begin{array}{ccc} t & g^+ & g^- \end{array}$$
$$U_1(l) = \mathrm{diag}(\eta \quad \tau \quad 1) \qquad\qquad (7.17)$$

It turns out that we must also take into account the d chirality:

$$\begin{array}{ccc} t & g^+ & g^- \end{array}$$
$$U_1(d) = \mathrm{diag}(\eta \quad 1 \quad \tau) \qquad\qquad (7.18)$$

Second-order steric interactions about the bonds flanking the β—carbons are appropriately defined in terms of meso and racemic dyads. These conformations are indicated below together with their statistical weights. These are *ll* dyads; the *dd* dyads (not shown) are their mirror images:

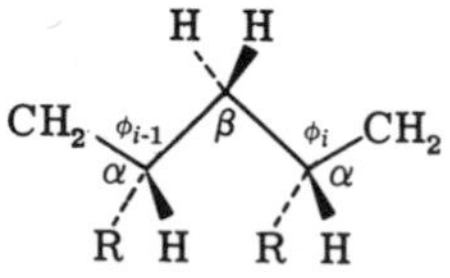

tt, $u_2 = \omega''$

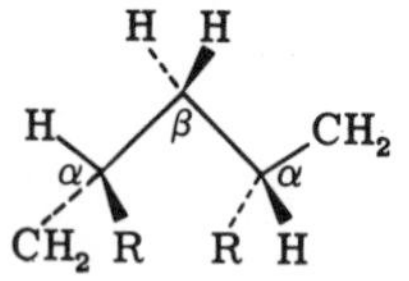

g^+t, $u_2 = \omega'$
↑↓
tg^-

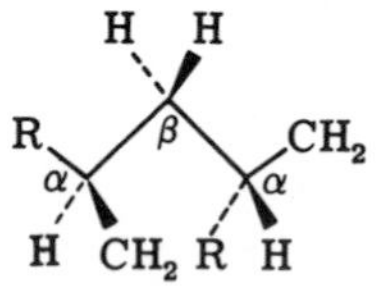

g^-t, $u_2 = 1$
↑↓
tg^+

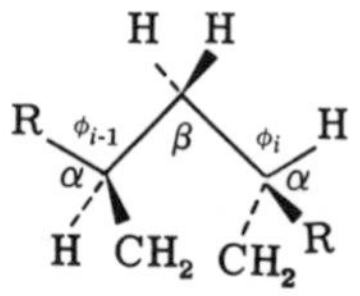

g^-g^-, $u_2 = \omega'$
↑↓
g^+g^+

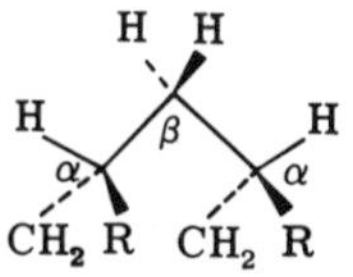

g^+g^-, $u_2 = \omega\omega''$

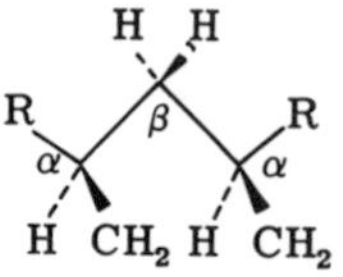

g^-g^+, $u_2 = \omega$

The statistical weight matrices incorporating both first- and second-order interactions are

$$U_{m_{12}}(ll) = \begin{array}{c} \\ t \\ g^+ \\ g^- \end{array} \begin{array}{c} \begin{array}{ccc} t & g^+ & g^- \end{array} \\ \begin{bmatrix} \eta\omega'' & 1 & \tau\omega' \\ \eta\omega' & \omega' & 0 \\ \eta & \omega & \tau\omega' \end{bmatrix} \end{array} \tag{7.19}$$

$$U_{m_{12}}(dd) = \begin{array}{c} \\ t \\ g^+ \\ g^- \end{array} \begin{array}{c} \begin{array}{ccc} t & g^+ & g^- \end{array} \\ \begin{bmatrix} \eta\omega'' & \tau\omega' & 1 \\ \eta & \tau\omega' & \omega \\ \eta\omega' & 0 & \omega' \end{bmatrix} \end{array} \tag{7.20}$$

The statistical weight of the g^+g^- conformation may be taken as 0.

For racemic dyads:

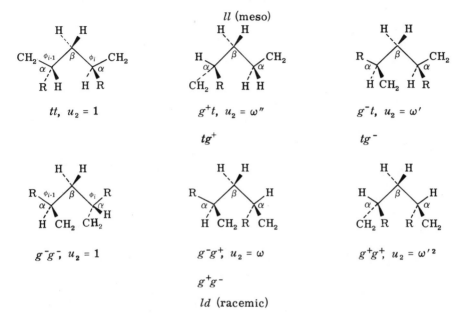

ll (meso)

tt, $u_2 = 1$

$g^+t, u_2 = \omega''$

tg^+

$g^-t, u_2 = \omega'$

tg^-

$g^-g^-, u_2 = 1$

$g^-g^+, u_2 = \omega$

$g^+g^+, u_2 = \omega'^2$

g^+g^-

ld (racemic)

The complete statistical weight matrices are

$$
U_{r_{12}}(ld) = \begin{array}{c} \\ t \\ g^+ \\ g^- \end{array}
\begin{array}{ccc} t & g^+ & g^- \end{array}
\left[\begin{array}{ccc}
\eta & \tau\omega'' & \omega' \\
\eta\omega'' & 0 & \omega \\
\eta\omega' & \tau\omega & 1
\end{array} \right]
\tag{7.21}
$$

$$
U_{r_{12}}(dl) = \begin{array}{c} \\ t \\ g^+ \\ g^- \end{array}
\begin{array}{ccc} t & g^+ & g^- \end{array}
\left[\begin{array}{ccc}
\eta & \omega' & \tau\omega \\
\eta\omega' & 1 & \tau\omega \\
\eta\omega'' & \omega & 0
\end{array} \right]
\tag{7.22}
$$

It is evident that here the $g^+g^+(ld)$ [and $g^-g^-(dl)$] conformations will be very strongly excluded and that the corresponding statistical weight may be assumed to be negligible whatever the nature of R.

The rotational states about the bonds flanking the α—carbons must also be taken into account (here we look from α toward β in defining the sense of rotation):

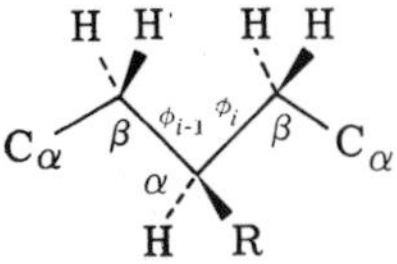

tt, $u_2 = 1$

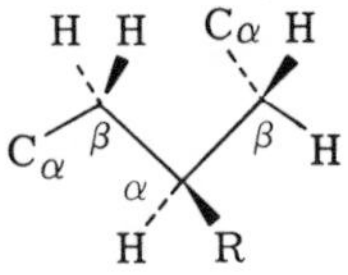

tg, $u_2 = 1$
↓↑
g^+t

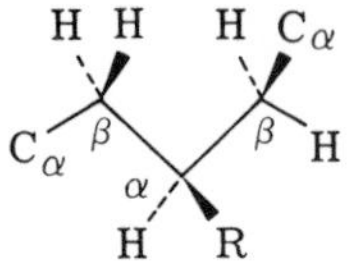

tg^-, $u_2 = 1$
↓↑
g^-t

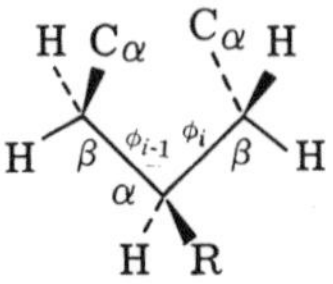

g^+g^+, $u_2 = 1$
↓↑
g^-g^-

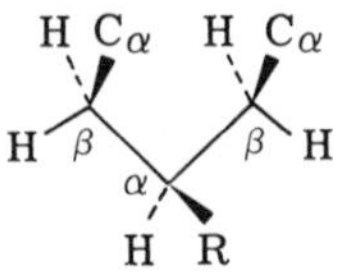

g^+g^-, $u_2 = \omega$

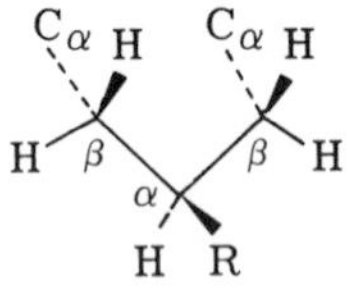

g^-g^+, $u_2 = \omega$

The $g^{\pm}g^{\pm}$ conformation involves no appreciable net interaction of the $C_\alpha s$ and their attached R groups, and are weighted as in the polymethylene chain (p. 192); but the g^+g^- and g^-g^+ conformations cause severe steric overlaps not only between the α—carbons but also between their attached R groups. These conformations may be assigned zero statistical weight unless the first atom in the R group is relatively small (F, OR′, OCOR′) or involves special interactions, such as the —O—H · · · O— hydrogen bonding in poly(vinyl alcohol.) The complete statistical weight matrices for the conformational states about α—carbons are accordingly

$$U_{\alpha_{12}}(d) = \begin{array}{c} \\ t \\ g^+ \\ g^- \end{array} \begin{array}{c} \begin{array}{ccc} t & g^+ & g^- \end{array} \\ \begin{bmatrix} \eta & 1 & \tau \\ \eta & 1 & \tau\omega \\ \eta & \omega & \tau \end{bmatrix} \end{array} \tag{7.23}$$

$$U_{\alpha_{12}}(l) = \begin{array}{c} \\ t \\ g^+ \\ g^- \end{array} \begin{array}{c} \begin{array}{ccc} t & g^+ & g^- \end{array} \\ \begin{bmatrix} \eta & \tau & 1 \\ \eta & \tau & \omega \\ \eta & \tau\omega & 1 \end{bmatrix} \end{array} \tag{7.24}$$

For calculations of the partition function and for other purposes, the matrices $U_{m_{12}}$ or $U_{r_{12}}$ are multiplied alternately in succession with the matrix $U_{\alpha_{12}}$. Chains may be generated corresponding to any value of P_m from isotactic ($P_m=1$) to syndiotactic ($P_m=0$), and their conformation-dependent properties can be calculated in the same manner as we have already described for simpler, symmetric chains.

The general conclusions from the consideration of the steric interactions in monosubstituted vinyl polymer chains are that *isotactic* chains tend to assume a conformation of alternating gauche (i.e., g^+ for d, g^- for l, the other gauche conformation being strongly excluded by $CH_2 \cdots C_\alpha H$ interactions) and trans bond rotational states. Such *(gt)(gt)* sequences, regularly repeated, generate a 3_1 helix, i.e., a helix that makes one turn for each three monomer units and consequently exhibits threefold symmetry viewed along the helical axis (see Fig. 7.7a). In solution, such helical sequences will be right-handed or left-handed with equal probability at any instant, but will reverse their chirality at random and with great rapidity, for the conformational states have very short lifetimes, as we have seen. The tendency of *(gt)(gt)* sequences to propagate themselves depends on the energies of the *tt* and $g^{\pm}g^{\mp}$ conformations, which of course in turn depend upon the nature of the $\alpha-$substituents. The junction between left- and right-handed helices represented by $\cdots (gt)(gt)(tg)(tg) \cdots$ (Fig. 7.4) is permitted to a substantial extent because the *tt* bond conformation at C_α (a) is energetically relatively favorable. However, the sequence $\cdots (tg)(tg)(gt)(gt) \cdots$ requires a $g^{\pm}g^{\mp}$ conformation (b), which involves $C_\alpha \cdots C_\alpha$ contacts, the energy of which, $E_{\omega''}$, is high. If the statistical weight of (b) is zero, only one junction of two helices of opposite sign is present; the chain cannot return again via conformation (a) to the other chirality. In fact, steric restrictions are not this stringent, and isotactic chains behave essentially as random coils.

For syndiotactic sequences, *tt* conformations are normally favored, with a substantial weight to $g^{\pm}g^{\pm}$, particularly in polyolefin chains, where dipole interactions are absent. In poly(vinyl chloride), for example, such interactions may be expected to favor *tt* conformations (Shimanouchi and Tasumi, 1961). For syndiotactic chains in solution, it is reasonable to expect substantial occurrence of helical conformations having sequences of the type $\cdots (gg)(tt)(gg)(tt) \cdots$. (As we shall see in Section 7.6, these expectations are generally borne out by the known crystalline conformations.)

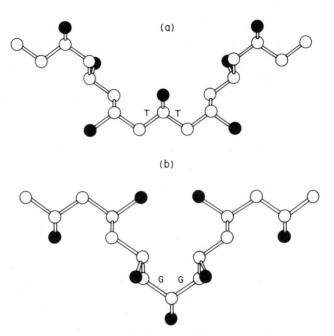

Fig. 7.4.   Conformational "defects" in 3_1 helices.

The successful calculation of conformation-dependent properties—unperturbed solution dimensions in particular—depends of course upon the proper choice of conformational states, i.e., rotational angles corresponding to energy minima, and upon the use of appropriate atom—atom or group—group potential energy functions, including electrostatic interactions in polar chains. Valence angles (not necessarily tetrahedral) and bond lengths must also be known. Many of these parameters may be fairly accurately estimated from X-ray and spectroscopic data on small molecules, but some remain adjustable within fairly wide limits; among these are E_η, E_ω, $E_{\omega'}$, and $E_{\omega''}$. In addition, the three-state model is itself an oversimplified one, although often adequate. (This can be seen even for n-pentane, in which there are actually energy minima near the high energy $g^\pm g^\mp$ state, achievable by appropriate adjustments of ϕ_2 and ϕ_3 to relieve the $CH_3 \cdots CH_3$ steric overlaps.) The result is that conformational calculations for most vinyl polymers—including α,α'-disubstituted chains, which we do not consider here—are semiempirical and must be accommodated to the experimental observations, usually by considerably complicating the relatively simple picture we have employed in this discussion.

As examples, let us briefly summarize the theoretical treatments of *polypropylene* and *polystyrene*. Calculations by Flory *et al.* (1966), employing a three-state model under the assumption that all the statistical weights ω, ω', and ω'' are zero, i.e., that conformations involving the four-bond CH_2 and side-chain interactions we have previously discussed are excluded, showed that the characteristic ratio C_∞, though somewhat sensitive to the allowed deviation of rotational angles, is relatively insensitive to the stereochemical configuration of the chain. It decreases somewhat from values of $10-12$ when $P_m = 0$ (i.e., a syndiotactic chain), to *ca.* 9 as P_m increases from 0.2 to 0.9 and then rises very rapidly to values exceeding 30 as P_m exceeds 0.95. Actual measurements of C_∞ for a polypropylene believed from proton NMR measurements to have an isotactic dyad content of ~ 0.98 yielded a value of 4.7 (Heatley *et al.*, 1969). Such high values of isotactic content are confirmed by ^{13}C NMR measurements discussed in Chapter 3. More recent calculations (Suter and Flory, 1975, however, using a five-state model with values of E_ω of ~ 1200 cal predict a monotonic decrease of C_∞ with isotactic dyad content to a minimum of less than 5 for a purely isotactic chain.

Similarly, earlier calculations (Flory *et al.*, 1966) of the dimensions of isotactic polystyrene chains indicated exclusion of the *tt* conformation owing to severe steric overlaps of the phenyl groups. More recent calculations (Yoon *et al.*, 1975) take into account the π-π attractive energy (~ 5 kcal) of the phenyl groups in this conformation. The absence of this attraction in other conformations is partially compensated for by solvent interactions; the net result is that the *tt* conformation in the meso configuration is believed to be permitted to an extent of *ca.* $8-12\%$. The calculated value of C_∞, excessively high in previous treatments, now agrees with the observed value of ~ 11.

7.5 EXPERIMENTAL OBSERVATION OF POLYMER CHAIN CONFORMATIONS IN SOLUTION

The detailed experimental observation of *local* polymer conformations in solution—as distinguished from end-to-end distances, dipole moments, and other similar "one-parameter" quantities—is not particularly easy, and has been accomplished principally by spectroscopic methods. Vibrational spectroscopy has had limited application, being mainly useful for biopolymers. There is ultraviolet spectroscopic evidence (Longworth, 1966; Vala and Rice, 1963) for a helical conformation of isotactic polystyrene. Probably the most powerful method is NMR, chiefly through exploitation of the strong angular dependence of the vicinal proton—proton J coupling. We cannot discuss the method in detail here, as it involves considerable complexities when long chains are involved. The observations are somewhat hampered by dipolar broadening. Interpretation of vinyl polymer spectra has been greatly aided by use of the small model compounds discussed in Chapter 3: meso and racemic 2,4-disubstituted pentanes and 2,4,6-trisubstituted heptanes having "isotactic", "heterotactic", and "syndiotactic" configurations. The results of such studies (see Bovey, 1972) in general support and confirm the steric influences and the results of the conformational calculations that we have just discussed.

The overall conformational state of polymer chains in solution, as distinguished from local conformations, has been traditionally studied by measurements of solution viscosity or radiation scattering, either of light or thermal neutrons. These methods have been supplemented by dipole moment and Kerr effect (electric birefringence), where applicable. It is not within the scope of this book to treat these matters in detail, and the reader is referred to other texts (Flory, 1953, 1969; Bovey and Winslow, 1979; Morawetz, 1975) and reviews (Higgins and Stein, 1978). It seems appropriate, however, to summarize briefly the use of solution viscosity measurements for the observation of end-to-end distances, as this is the commonest method for determining the characteristic ratio C_∞. The *intrinsic viscosity* is defined by:

$$[\eta] = \lim \left[\frac{\eta - \eta_0}{\eta_0}\right]_{c_2 \to 0} \Big/ c_2 \qquad (7.25)$$

where c_2 is the polymer concentration and η and η_0 are the polymer solution and solvent viscosities, respectively. The customary units of c_2 are grams per deciliter, and so $[\eta]$ is expressed in dL-g^{-1} The intrinsic viscosity may be expressed as (Flory, 1953):

$$[\eta] = \frac{\phi <r^2>^{3/2}}{M} \qquad (7.26)$$

where ϕ is a universal constant equal to approximately 2.5×10^{21} and M is the molecular weight. The quantity $<r^2>$ is larger than $<r^2>_0$ because of expansion of the molecular coil in most real solvents. The expansion factor is designated α, so that

$$<r^2> = \alpha^2 <r^2>_0 \qquad (7.27)$$

In a θ solvent, the expanding effect of *excluded volume* is just balanced by the contracting effect of polymer segment—segment interaction and $\alpha = 1$; the chain has its unperturbed dimensions, and consequently

$$C_\infty = \frac{<r^2>_0}{nl^2} = \frac{1}{nl^2}\left[\frac{[\eta]_\theta M}{\phi}\right]^{2/3} \qquad (7.28a)$$

where $n = M/m$, m being the monomer unit molecular weight. Thus

$$C_\infty = \frac{m}{Ml^2}\left[\frac{[\eta]_\theta M}{\phi}\right]^{2/3} \qquad (7.28b)$$

where l is usually about 0.15 nm. M can be obtained from light scattering, gel permeation chromatography, or viscosity measurements. It is preferable that the molecular weight distribution be narrow.

In Table 7.3 a few selected values of C_∞ (that is, $<r^2>_0/nl^2$) are shown (see Flory, 1969). The value of *ca.* 2.0 for polytetramethylene

TABLE 7.3

Characteristic Ratios C_∞ for Selected Polymer Chains

Polymer	Conditions	$C_\infty = <r^2>_0/nl^2$
Polymethylene	Decanol-1, 138°	6.7
Polyoxymethylene	—	*ca.* 10
Polyoxyethylene	Aqueous K_2SO_4, 35°	4.0
Polytetramethylene oxide	—	*ca.* 2.0
Polypropylene isotactic atactic	Diphenyl ether 145° 153°	4.7 (5.3)
Polystyrene atactic	Diphenyl ether Cyclohexane, 35°	10.2
Poly(methyl methacrylate) atactic isotactic	Various solvents Acetonitrile, 28°	6.9 9.3
Polydimethylsiloxane	Butanone	6.2

oxide should not be taken as indicating that this chain is freely rotating (Section 7.3), as this unusually collapsed conformation is the result of a cancellation of effects. As the succession of methylene groups is increased beyond four, C_∞ increases again; it clearly must approach the value of 6.7 characteristic of polymethylene.

7.6 EXPERIMENTAL OBSERVATION OF POLYMER CHAIN CONFORMATIONS IN THE CRYSTALLINE STATE

In the *solid state*, vibrational spectroscopy has proved to be of considerable power for the determination of chain conformations, particularly in crystallizable polymers. The splitting of the CH_2 rocking band reflects the interaction of the chains with their neighbors in the crystalline array and is consistent with a planar zigzag conformation. By the method of *normal coordinate analysis*, it is possible to account for the positions and splittings of the vibrational lines in IR and Raman spectra and thereby to establish the intrachain and interchain force field. This has been particularly successful for polyethylene (Tasumi and Krimm, 1967) and for stereoregular vinyl polymers (Miyazawa, 1967).

The primary method for the determination of macromolecular conformations in the crystalline solid state is *X-ray diffraction*. Valuable contributions have also been made using *electron diffraction*. Space does not permit us to discuss the basis of the method here. For this there are many excellent sources; some of these are listed under "X-Ray Diffraction" in the General References. We shall deal only with the results.

The influences that determine the conformation of a polymer chain in the crystalline state are primarily those that we have already described as determining the solution conformation. In addition to these intramolecular requirements, one must also consider intermolecular energies associated with chain packing. These are generally relatively small [except in cases where interchain hydrogen bonding is involved, as in polyamides (p. 213)] but may affect the choice between conformations having nearly equal internal energies and will play a part in first-order transitions between crystalline forms.

We have seen that for *polyethylene* the trans−trans or planar zigzag conformation is the form of lowest energy and that the vibrational spectra are consistent with this finding. The crystal structure is shown in Fig. 7.5 (Bunn, 1946). The repeat distance along the chain axis is 0.253 nm, and represents the length of one monomer unit, i.e., twice the projection of the C−C bond distance on the axis:

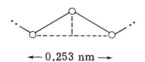

←— 0.253 nm —→

The portion of the structure that reproduces the whole when repeated in three dimensions is the *unit cell* and is represented by the parallelopiped at the bottom of the figure. It contains segments of two chains (shown in the center; for clarity, neighboring chains are reproduced in lighter tones) and is the same as that of crystalline linear paraffins. The morphology of polyethylene is actually considerably more complex than Fig. 7.5 suggests owing to the occurrence of amorphous regions, lattice defects, and folds. The reader is referred to Bovey and Winslow (1979, Chapter 5) for further discussion of these matters.

The substitution of hydrogen by the somewhat larger fluorine atom increases the internal energy of the trans—trans conformation enough so that the chains of *polytetrafluoroethylene* exhibit instead a twisted structure. The three-bond rotational angle is increased from 0° to 14° to minimize steric interactions, and this slow-turning ribbon consequently repeats every 13 carbon atoms (1.69 nm), constituting a 13_6 helix, corresponding to 2.17 monomers per turn. Unlike those of

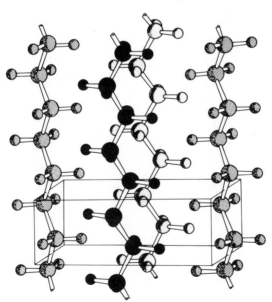

Fig. 7.5.   The crystal structure of polyethylene (from Bunn, 1946).

polyethylene, the chains of polytetrafluoroethylene are nearly cylindrical in cross section. The structure at a temperature below approximately 19° is shown in Fig. 7.6. Above 19°, the conformation undergoes a first-order crystalline transformation in which the chains exhibit a 15_7 helical conformation with a torsional angle of about 12° and a repeat distance of 1.95 nm. Above 30°, the structure becomes a torsionally oscillating one.

We have seen (Section 7.4) that in the chains of *polyoxymethylene*, unlike those of polyethylene, the gauche conformation is of lower energy than the trans. The most stable crystalline chain conformation is in consequence a 9_5 helix (Fig. 7.7). The repeat distance is

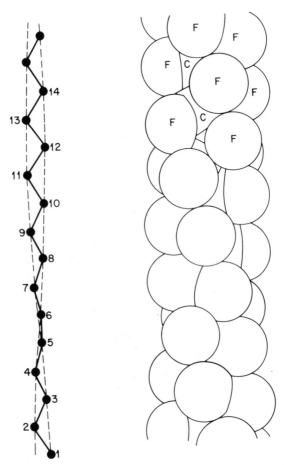

Fig. 7.6.   Conformation of polytetrafluoroethylene (from Bunn and Howells, 1954).

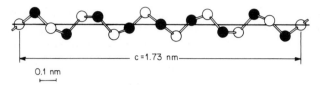

Fig. 7.7. The conformation of polyoxymethylene. Black circles represent carbon atoms and white circles oxygen atoms (from Uchida and Tadokoro, 1967).

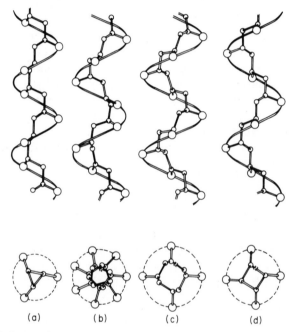

Fig. 7.8. Helical conformations of isotactic polymers with varying side-chains (from Natta and Corradini, 1960).

(a) R = $-CH_3$, $-C_2H_5$, $-CH=CH_2$, $-CH_2CH_2CH(CH_3)_2$, $-OCH_3$, $-OCH_2CH(CH_3)_2$

(b) R = $-CH_2CH(CH_3)CH_2CH_3$, $-CH_2CH(CH_3)_2$

(c) R = $-CH(CH_3)_2$, $-C_2H_5$

(d) R =

1.73 nm. The COC bond angle is 112° and the OCO bond angle 111°. The helices interlock tightly and are all of one chirality in a single crystal or in large portions of it (Uchida and Tadokoro, 1967).

We have seen that in isotactic vinyl polymer chains the preferred conformation is often the 3_1 helix, generated from alternating gauche and trans conformations. This is true for polypropylene, polybutene-1, poly(5-methylhexene) ($R = -CH_2CH_2CH(CH_3)_2$), poly(vinyl methyl ether), poly(vinyl isobutyl ether) ($R = -OCH_2CH(CH_3)_2$), and polystyrene (Fig. 7.8a). If the side-chain is bulkier, particularly near its attachment to the α-carbon, the helix expands; thus poly(4-methylhexene-1) ($R = -CH_2CH(CH_3)CH_2CH_3$) and poly(4-methyl-pentene-1) ($R = -CH_2CH(CH_3)_2$) exhibit 7_2 helices (Fig. 7.8b) while poly(3-methylbutene-1) ($R = i$-propyl), poly(o-methylstyrene), poly(o-methyl-p-fluorostyrene), and poly(α-vinylnaphthalene) have 4_1 helices. Surprisingly, poly(o-fluorostyrene) crystallizes with a 3_1 helical conformation, whereas with the fluorine in the para position a 4_1 helix is preferred. This is an instance of chain conformation being dictated by crystalline packing requirements.

A number of isotactic vinyl polymers show *polymorphism*. Polypropylene has been reported to exist in four crystalline modifications, all 3_1 helices but differing in the details of the packing of the chains. Polybutene-1 initially crystallizes from the melt as an 11_3 helix, which then slowly transforms to a room-temperature-stable 3_1 helical form; the equilibrium transition between them occurs at about 110° (J. P. Luongo, private communication, 1975).

The packing of vinyl chains having the same conformation may occur in several ways owing to the fact that the helices have a *sense of direction* (even though the basic chain structure does not) as well as chirality. This is illustrated in Fig. 7.9, which shows a side view of a right-handed 3_1 helix of isotactic polystyrene. In common with all such helices, the α-substituents make an angle with the helical axis. (In the case of polystyrene, the orientation of the plane of the phenyl group is such as to bisect the $C_\beta C_\alpha C_\beta$ angle.) Packing energies may differ depending upon whether each chain is surrounded by other chains—usually with a coordination of 3—of the same or of opposite direction, and of the same or of opposite chirality. In polystyrene and polypropylene, it appears that alternating chirality prevails, but it is not obvious whether the directions of the helices are (a) entirely random or (b) uniform over small domains or even whole single crystals. It is probable that in the most stable form of polypropylene, (b) is the case.

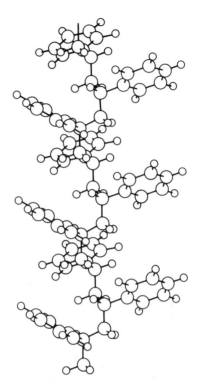

Fig. 7.9.   Side view of the 3_1 helix of isotactic polystyrene (from Natta *et al.*, 1960).

There is relatively little information concerning the conformations of syndiotactic vinyl polymer chains in the crystalline state owing to the scarcity of authentically established syndiotactic polymers. Syndiotactic *poly(1,2-butadiene)* (Natta and Corradini, 1956) and *poly(vinyl chloride)* (Natta and Corradini, 1956) crystallize in a planar zigzag conformation. Syndiotactic *polypropylene* forms a $\cdots (gg)(tt)(gg)(tt) \cdots$ helix. The side and end views, together with the crystal packing, are shown in Figs. 7.10 and 7.11. There are four monomer units per turn (Natta *et al.*, 1960). It is of interest that the all-*trans* planar zigzag crystalline conformation can also be prepared by quenching in ice-water and stretching but that it is metastable and transforms into the *(gg)(tt)* helix at room temperature (Natta *et al.* 1964).

Of the many $\alpha,\alpha-$disubstituted vinyl polymer chains, *polyisobutylene* is of particular interest. This polymer is rubbery in the normal solid state, but the structural regularity of the chains permits a considerable degree of crystallinity to develop in stretched fibers. Severe steric overlap of the methyl groups rules out a planar zigzag

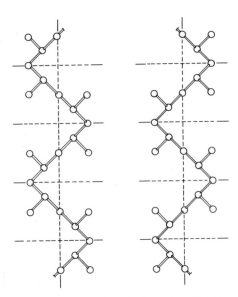

Fig. 7.10. Side view of enantiomorphous helices of syndiotatic polypropylene (from Natta *et al.*, 1960).

model. The conformation in best agreement with X-ray (Fuller *et al.*, 1940; Liquori, 1953, 1955; Bunn and Holmes, 1958) and energy calculations (Allegra *et al.*, 1970) appears to be an 8_3 helix (Fig. 7.12).

Natural rubber [poly(*cis*-1, 4-isoprene)] also crystallizes on stretching. This structure, one of the earliest determined by X-ray for

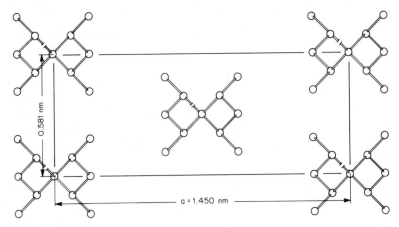

Fig. 7.11. Mode of packing of syndiotactic polypropylene helices (from Natta *et al.*, 1960).

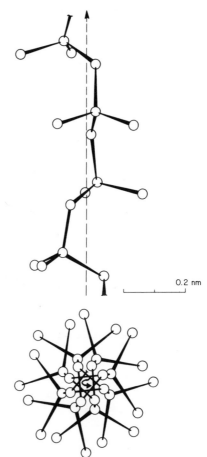

Fig. 7.12. Crystalline conformation in stretched fibers of polyisobutylene (from Allegra *et al.*, 1970).

0.2 nm

a macromolecule, is shown in Fig. 7.13. The repeat distance along the chain is 0.81 nm, indicating two isoprene units per conformational repeat (Bunn, 1942).

The chain conformation and crystal morphology of *polyamides* are strongly influenced by the presence of hydrogen bonds between the amide carbonyl and NH groups of neighboring chains, resulting in the existence of sheets. The forces stabilizing the sheets are considerably greater than those between them. The chains are planar zigzags and in nylon 66 are packed as shown in Fig. 7.14a (Bunn and Garner, 1947), which represents the crystalline form designated as α. There are also β and λ forms that have the same structure within each sheet but differ somewhat in the offset arrangement of the sheets with respect to each other.

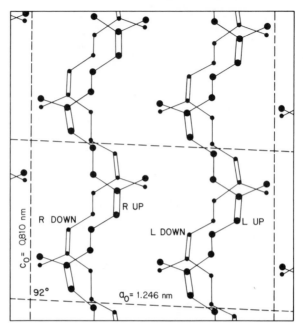

Fig. 7.13. The chain conformation and crystal packing in stretched natural rubber (from Bunn, 1942).

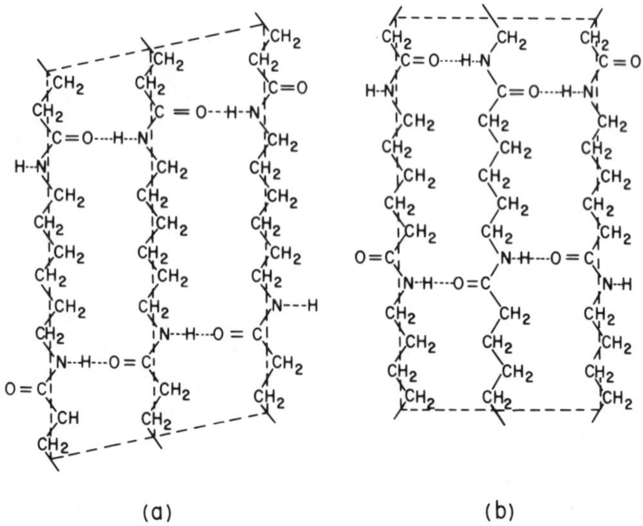

(a)   (b)

Fig. 7.14. Chain conformations of polyamides: (a) nylon 66; (b) nylon 6 (from Holmes *et al.*, 1955).

In nylon 6, unlike nylon 66, the chains have a structural sense of direction and hence may in principle be arranged parallel or antiparallel in the hydrogen bonded sheets. Both models and X-ray data indicate an antiparallel arrangement, in which, as shown in Fig. 7.14b, all hydrogen bonds can be readily formed (Holmes *et al.*, 1955). The best defined form is again termed α; in this form, the arrangement of sheets is the same as in the α form of nylon 66.

7.7 EXPERIMENTAL OBSERVATION OF POLYMER CHAIN CONFORMATIONS IN THE AMORPHOUS SOLID STATE

For some years there has been a question concerning the conformations of macromolecules in the bulk amorphous state, both rubbery and glassy. One school of thought has maintained that the polymer chains coil upon themselves and do not mingle with their neighbors, whereas another school has maintained that the molecules do quite the opposite, responding to their neighbors by forming bundles of chains that are correlated to a substantial extent, though falling short of the degree of order detectable by X-ray diffraction [for reviews, see Tonelli, 1977; Flory, 1953; see also "Organization of Macromolecules in the Condensed Phase," Faraday Discussions No. 68, (1979)]. Flory, (1953) postulated long ago that in fact they do neither, but instead exhibit the unperturbed dimensions characteristic of their behavior in θ solvents, i.e., the dimensions discussed in Section 7.4. It will be recalled that in θ solvents the expanding tendency of the chain due to excluded volume effects is just compensated by the contracting effect of attractive polymer segment—segment interactions, so that unperturbed dimensions are attained. A polymer molecule surrounded by other like polymer molecules obviously experiences interference with itself, i.e., an excluded volume effect. But it has nothing to gain by expanding, for the decreased interaction with itself thereby attained would be just compensated by increased interaction with its neighbors. Its conformation should therefore be unperturbed, although strictly speaking a polymer dissolved in itself is not under θ conditions since the thermodynamic interaction parameter is zero, which is not the case in a θ solvent.

The controversy concerning the dimensions of polymer molecules in the amorphous solid state has been kept alive by the absence until recently of any means of directly measuring the mean square end-to-end distances or radii of gyration of polymers in this state. This lack has now been supplied by *small angle coherent neutron scattering*. This technique uses a beam of "cold" neutrons (from a reactor) having a deBroglie wavelength h/mv of about 0.5 nm, corresponding to a particle velocity of *ca.* 1000 m-s^{-1} or an energy of about 70 cal/Einstein or 23 cm^{-1}. This beam is employed in a small-angle scattering technique, which in principle entirely parallels light scattering (Allen, 1976; Higgins and Stein, 1978). The coherent scattering of neutrons is about three times as great from deuterons as from protons, and this supplies a parallel to the difference in refractive index between polymer and solvent in Rayleigh scattering. For example, one may prepare a solution of $1-2\%$ normal poly(methyl methacrylate) in perdeuterated monomer, $CD_2=C(CD_3)COOCD_3$ and polymerize the latter, giving a solid sample with proton polymer molecules dispersed in deuterated polymer. (One may also employ a deuterated polymer in a matrix of protonated polymer.) Some results for poly(methyl methacrylate) are shown in Table 7.4 (Kirste *et al.*, 1973). This polymer was nearly monodisperse and had a molecular weight of 250,000.

TABLE 7.4

Neutron Scattering Results for Poly(methyl methacrylate)

Conditions	$<s^2>^{1/2}$, nm*	$<r^2>$, nm^2	$<r^2>/nl^2$
Amorphous solid	12.5	938	8.3
In *n*-butyl chloride, θ solvent	11.0	726	6.5
In dioxane, "good solvent"	17.0	1734	15.4

*Recall that for long polymer chains under unperturbed conditions:

$$<s^2>_0 = \frac{1}{6} <r^2>_0.$$

For this polymer, DP $\simeq$ 2500; $n = 5000$; $l = 0.15$ nm; $nl^2 = \sim 112$.

Thus, Flory's postulate is at least approximately confirmed. Further neutron scattering data (Benoit *et al.*, 1973) for three fractions of atactic polystyrene in bulk and in cyclohexane under θ conditions are presented in Table 7.5.

These results further confirm the unperturbed dimensions of an amorphous bulk polymer, and also support the theoretical expectation that the characteristic ratio should be independent of molecular weight. The characteristic ratio of polystyrene from viscosity measurements (Table 7.3) agrees with that obtained by neutron scattering.

TABLE 7.5

Neutron Scattering Results for Polystyrene

$\overline{M}_w$	bulk			in cyclohexane		
	$<s^2>^{1/2}$,nm	$<r^2>$,nm^2	$<r^2>/nl^2$	$<s^2>^{1/2}$, nm	$<r^2>$,nm^2	$<r^2>/nl^2$
21,000	3.8	86.6	9.6	4.2	106	11.6
325,000	14.3	1227	8.8	15.0	1357	9.6
1,100,000	29.7	5300	11.1	29.3	5150	10.8

7.8   CHAIN CONFORMATION AND CHEMICAL SHIFT

We have seen in Chapter 3 that NMR chemical shifts, particularly of ^{13}C, are highly sensitive to the stereochemical configuration of vinyl polymer chains. It has been postulated (Grant and Cheney, 1967; Carman, 1973; Tonelli and Schilling, 1981) that two carbon atoms separated by three intervening bonds shield each other by *ca.* 5.0 ppm

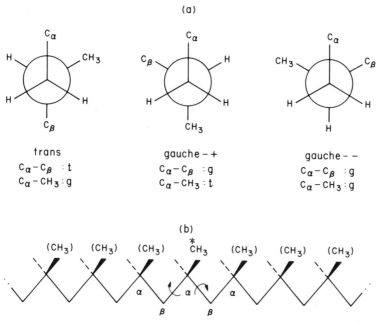

Fig. 7.15.   (a) Conformations of a four-carbon fragment of a polypropylene chain (*d* configuration); (b) heptad of polypropylene chain; observed methyl is marked by *

when the central bond is gauche compared to their chemical shifts when it is trans. These conformations are illustrated in Fig. 7.15 for a four-carbon fragment of a polypropylene chain. It can be seen that when the main chain is trans, the C_α and CH_3 carbons are in a gauche relationship; when the chain is gauche+ the C_α and C_β carbons are of course gauche. When the chain is gauche− (a high energy state that, as we have seen, is nearly excluded) the C_α experiences gauche interactions with both the CH_3 and the C_β carbon. The CH_3 chemical shift is then sensitive to the gauche content of the main chain bonds flanking the α−carbon to which it is attached. This is illustrated in Fig. 7.15b, which represents a configurational heptad sequence of a polypropylene chain. Not only triad but pentad and even heptad sequences may be resolved in polypropylene spectra. In Fig. 7.16 is shown the 90 MHz ^{13}C spectrum of an atactic polypropylene, observed in 1,2,4-trichlorobenzene solution at 100° (Tonelli and Schilling, private communication, 1981). We learned in Chapter 3 that a maximum of 36 possible heptad sequences may occur in such a polymer. Below the experimental spectrum is a line spectrum in which the expected chemical shifts of the central methyl group carbons of the 36 heptads shown above the spectrum are indicated by number. These

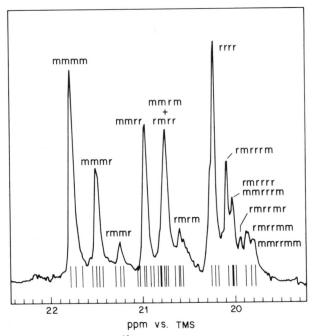

Fig. 7.16. The 90 MHz methyl ^{13}C spectrum of atactic polypropylene, observed in 1,2,4-trichlorobenzene at 100°.

chemical shifts have been calculated by the rotational isomeric state method (Tonelli and Schilling, 1981) using the five-state conformational model of Suter and Flory (1975) to estimate the gauche content of the central bonds of each heptad. Consider, for example, the heptads *mmmmmm* and *mmmmmr*. The discrimination in chemical shift between the central methyl groups in these sequences arises from the inversion of the chirality of one $\alpha-$carbon seven bonds removed from the observed carbon. It is believed that this influence is propagated down the chain not in some mysterious way but by an influence on the gauche content of successive bonds. The fact that the line spectrum matches closely the observed spectrum shows that the gauche hypothesis must be essentially correct and that the Suter—Flory treatment provides an accurate estimate of the chain conformation as a function of stereochemical configuration. The ^{13}C spectra of the $\alpha-$ and $\beta-$carbons are predicted with comparable accuracy, as is the methyl carbon spectrum of the model heptadecanes discussed in (Section 3.6.1). Extension to other polymers such as poly(vinyl chloride) and polystyrene has met with considerable success, although it is evident that in addition to conformational influences there are specific solvent effects not related to conformation.

In Chapter 8 we shall see that the gauche model may also be invoked to interpret the ^{13}C NMR spectra of polymers in the *solid state*.

REFERENCES

Abe, A., Jernigan, R. L., and Flory, P. J. (1966). *J. Am. Chem. Soc.* **88**, 631.
Allegra, G., Benedetti, E., and Pedone, C. (1970). *Macromolecules* **3**, 727.
Allen, G. (1976). *In* "Structural Studies of Macromolecules by Spectroscopic Methods" (K. J. Ivin, ed.). Wiley, New York.
Benoit, H., Decker, D., Higgins, J. S., Picot, C., Cotton, J. P., Farnoux, B., Jannink, G., and Ober, R. (1973). *Nature (London)* **245**, 13.
Bovey, F. A. (1972). "High Resolution NMR of Macromolecules." Academic Press, New York.
Bovey, F. A., and Winslow, F. H. (1979). "Macromolecules." Academic Press, New York.
Bunn, C. W. (1942). *Proc. Roy. Soc. London, Ser. A*, **180**, 40.
Bunn, C. W. (1946). "Chemical Crystallography." Oxford Univ. Press, London and New York.
Bunn, C. W., and Garner, E. V. (1947). *Proc. Roy. Soc. London, Ser. A*, **189**, 39.
Bunn, C. W., and Holmes, D. R. (1958). *Discuss. Faraday Soc.* **25**, 95.
Bunn, C. W., and Howells, E. R. (1954). *Nature (London)* **174**, 549.
Carman, C. J. (1973). *Macromolecules* **6**, 725.

Flory, P. J., Mark, J. E., and Abe, A. (1966). *J. Am. Chem. Soc.* **88**, 639.
Fuller, C. S., Frusch, C. J., and Pape, N. R. (1940). *J. Am. Chem. Soc.* **62**, 1905.
Grant, D. M., and Cheney, B. V. (1967). *J. Am. Chem. Soc.* **89**, 5315, 5319.
Heatley, F., Salovey, R., and Bovey, F. A. (1969). *Macromolecules* **2**, 619.
Higgins, J. S., and Stein, R. S. (1978). *J. Appl. Crystallogr.* **11**, 346.
Holmes, D. R., Bunn, C. W., and Smith, D. J. (1955). *J. Polymer Sci.* **17**, 159.
Kirste, R. G., Fruse, W. A., and Schelten, J. (1973). *Makromol. Chem.* **162**, 299.
Liquori, A. M. (1953). *Proc. 13th IUPAC Meeting, Stockholm.*
Liquori, A. M. (1955). *Acta Crystallogr.* **9**, 345.
Longworth, J. W. (1966). *Biopolymers* **4**, 1131.
Mark, J. E., and Flory, P. J. (1965). *J. Am. Chem. Soc.* **87**, 1415.
Mark, J. E., and Flory, P. J. (1966). *J. Am. Chem. Soc.* **88**, 3702.
Miyazawa, T. (1967). *In* "The Stereochemistry of Macromolecules" (A. D. Ketley, ed.), Vol. 3, Chap. 3. Marcel Dekker, New York.
Morawetz, M. (1975). "Macromolecules in Solution." Wiley, New York.
Natta, G., and Corradini, P. (1956). *J. Polymer Sci.* **20**, 251.
Natta, G., and Corradini, P. (1960). *Nuovo Cim., Suppl. 15* **1**, 9.
Natta, G., Corradini, P., and Ganis, P. (1960). *Makromol. Chem.* **39**, 238.
Natta, G., Peraldo, M., and Allegra, G. (1964). *Makromol. Chem.* **75**, 215.
Shimanouchi, T., and Tasumi, M. (1961). *Spectrochim. Acta* **17**, 775.
Suter, U. W., and Flory, P. J. (1975). *Macromolecules* **8**, 765.
Tasumi, M., and Krimm, S. (1967). *J. Chem. Phys.* **39**, 2348.
Tonelli, A. E., and Schilling, F. C. (1981). *Acc. Chem. Res.* **14** 233.
Uchida, T., and Tadokoro, H. (1967). *J. Polymer Sci., Part A-2* **5**, 63.
Vala, M. T., Jr., and Rice, S. A. (1963). *J. Chem. Phys.* **39**, 2348.
Yoon, D. Y., Sundararajan, P. R., and Flory, P. J. (1975), *Macromolecules* **8**, 776.

GENERAL REFERENCES

I Conformations of Polymer Chains

Birshtein, T. M., and Ptitsyn. O. B. (1966). "Conformations of Macromolecules" (translated from the Russian by S. N. Timasheff and M. J. Timasheff). Wiley (Interscience), New York.
Flory, P. J. (1953), "Principles of Polymer Chemistry." Cornell Univ. Press, Ithaca, New York.
Flory, P. J. (1969). "Statistical Mechanics of Chain Molecules." Wiley (Interscience), New York.
Tonelli, A. E. (1977). Polymer conformation and configuration. *In* "Encyclopedia of Polymer Science and Technology" (N. Bikales, ed.), 2nd ed. Wiley, New York.
Vol'kenshtein, M. V. (1963). "Configurational Statistics of Polymer Chains" (translated from the Russian by S. N. Timasheff and M. J. Timasheff). Wiley (Interscience), New York.

II X-Ray Diffraction and Polymer Chain Conformations in the Crystalline State

Corradini, P. (1968). "The Stereochemistry of Macromolecules," Vol. 3, Chapter 1. Marcel Dekker, New York.
Geil, P. H. (1963). "Polymer Single Crystals." Wiley (Interscience), New York.
Tadokoro, H. (1979). "Structure of Crystalline Polymers." Wiley, New York.
Wunderlich, B. (1973). "Macromolecular Physics," Vol. 1. Academic Press, New York.

Chapter 8

SOLID STATE NMR
OF MACROMOLECULES

L. W. Jelinski

8.1 INTRODUCTION

Schaefer and Stejskal in 1976 reported the first truly high resolution NMR spectra of solid polymers (Schaefer and Stejskal, 1976). Their approach involved the combination of three already-known techniques: dipolar decoupling, cross polarization, and magic angle spinning. Since that time much progress has been made in the field of solid state NMR, and now commercial spectrometers are available that perform most solid state NMR experiments. This progress is due in part to the development of pulsed Fourier transform NMR techniques.

8.2 PULSED FOURIER TRANSFORM NMR AND RELAXATION

As we have seen in Chapter 2, the NMR phenomenon is described as a process wherein the normally degenerate nuclear spin energy levels lose their degeneracy in the presence of a magnetic field. When the magnetic moment μ of the nucleus interacts with the magnetic field B_0, the torque exerted on the spinning nucleus causes the nuclear moment to precess about B_0. The *Larmor frequency* ω_0 is the frequency of this precession, and is related both to the gyromagnetic ratio γ of the nucleus and to the magnetic field by $\omega_0 = \gamma B_0$. As Fig. 2.13 illustrated, the separation of these energy levels depends upon the magnetic field strength, and can be expressed as

$$\frac{\gamma h B_0}{2\pi} \tag{8.1}$$

where γ is the gyromagnetic ratio of the nucleus under consideration. More of the spins tend to align their magnetic moments along the field direction than against it, with the population of spins in each energy level (N_+ and N_-) defined by the Boltzmann distribution

$$\frac{N_-}{N_+} = e^{-\Delta E/kT} \tag{8.2}$$

where N_- is the population of the upper energy level. Transitions between these energy levels form the basis of NMR spectroscopy.

The Boltzmann distribution of spins is not established instantaneously when the sample is placed in the magnetic field. Instead, equilibration of the spin populations is a time dependent process involving the interaction of the nuclei with their surroundings, or *lattice*. The *spin lattice relaxation time* T_1 defines the rate of this equilibration process. (The interested reader is referred to the references at the end of Chapter 2 for a more complete discussion of nuclear relaxation. It is our intention here only to introduce the concept of spin—lattice relaxation, as it will appear again in our discussion of solid state NMR techniques.)

Transitions between the nuclear spin energy levels can be induced by bringing each chemically shifted nucleus in the sample into resonance. (As we saw in Chapter 2, the range of frequencies for a particular nucleus is actually very small; for ^{13}C it is of the order of several hundred parts per million.) The resonance condition is accomplished by one of three methods: either the field is swept, a range of frequencies is swept, or a band of frequencies about the Larmor frequency is excited simultaneously. Fourier transform (FT) NMR employs the third method, where a short rf pulse, at the Larmor frequency, is applied to the system. The pulse is generally powerful enough to cover the entire frequency range of the nuclei in the sample. The Boltzmann distribution of spin populations is thus disturbed from equilibrium, and the system strives to reestablish its equilibrium state through transitions between the energy levels. The return to equilibrium is again a time dependent process involving both spin−lattice (T_1) and spin−spin (T_2) interactions. Following the rf pulse, a radio receiver is turned on to receive the signal from the nuclei. This signal, recorded as a function of time, is called the *free induction decay* (FID) (Fig. 8.1a).

Functions such as the FID in Fig. 8.1a can be expressed as an infinite sum of sine and cosine functions, called a Fourier series. The time domain function *f(t)* of Fig. 8.1(a) and the frequency domain function $F(\omega)$ in Fig. 8.1b are Fourier inverses. Thus, the frequency domain spectrum in Fig. 8.1b is obtained by

$$F(\omega) = \int_{-\infty}^{\infty} f(t)\, e^{-i\omega t}\, dt \qquad (8.3)$$

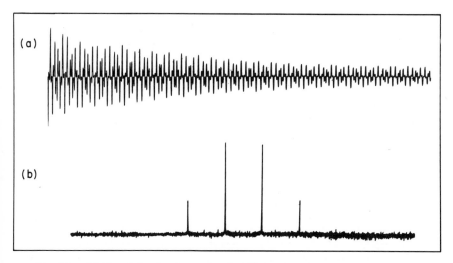

Fig. 8.1. (a) Free induction decay signal of ^{13}C-enriched methyl iodide in the time domain and (b) the corresponding frequency domain spectrum after Fourier transformation of the signal in (a).

In practice, the function $f(t)$ is evaluated using the Cooley—Tukey "fast Fourier transform" (Cooley and Tukey, 1965; Cochran et al., 1967), an efficient algorithm that has been widely applied in modern minicomputers.

It is often convenient to conceptualize FT NMR from the viewpoint of a rotating frame of reference. (Earth is a familiar rotating frame of reference with a 24-hr period.) The behavior of spins can be described by simple vector pictures in a coordinate system rotating at the Larmor frequency. The equilibrium distribution of spins can be described as a net magnetization (M_0) in the field direction (Fig. 8.2a). The effect of an rf field B_1, applied perpendicular to the magnetic field direction, is shown in Fig. 8.2b. The B_1 field exerts a torque on the macroscopic magnetization, causing M_0 to precess in the $x'y'$ plane at a rate $\omega = \gamma B_1$. The duration of the B_1 pulse is sufficient to tip the

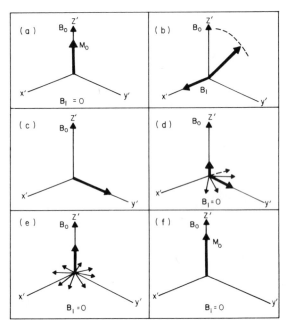

Fig. 8.2. Rotating frame diagrams describing the pulsed NMR experiment. (The "primed" axis is one used to indicate a rotating coordinate system.) (a) The net magnetization M_0 is aligned along the magnetic field direction; (b, c) an rf field B_1 is applied perpendicular to B_0. The duration of B_1 is sufficient to tip the net magnetization by 90°; (d, e) the spins begin to relax in the $x'y'$ plane by spin—spin (T_2) processes and in the z' direction by spin—lattice (T_1) processes; (f) the equilibrium magnetization is reestablished along B_0.

magnetization by 90° into the $x'y'$ plane (Fig. 8.2c), where the detection coil is located. In Figs. 8.2d and 8.2e the spins are seen relaxing back to their equilibrium state by T_1 and T_2 relaxation processes. The return to equilibrium is shown in Fig. 8.2f. The experiment can be repeated once equilibrium is reestablished. Thus, relaxation times control the length of time required to obtain an FT NMR spectrum. Carbon spin−lattice relaxation times for solids are often very long (several minutes), whereas T_1 values for some polymers in solution are of the order of tenths of seconds. (We shall return to the problem of long carbon T_1s in solids when we discuss cross polarization in Section 8.7.)

The advantages of pulsed Fourier transform techniques are primarily related to time savings. The data can be collected all at once, rather than from a slow sweep of the field. In addition, FT methods are useful for signal averaging. Many free induction decays from a weak signal can be repeatedly added in the minicomputer. The noise is random and cancels out, but the signal is coherent and continually adds. The signal-to-noise ratio increases as the square root of the number of accumulated spectra.

8.3 DIPOLAR BROADENING IN SOLIDS

One generally does not encounter the static dipole−dipole interaction in NMR spectra of small molecules in solution, as rapid molecular tumbling averages this interaction to zero. In solids, however, the molecules are not free to tumble isotropically and the static dipole−dipole interaction generally imposes a broadening so large that all spectral details are lost.

The dipole−dipole interaction arises from *direct* spin−spin coupling. That is, the magnetic dipole of one spin (^{1}H, for example) influences the magnetic dipole of another spin (^{13}C, for example). This results in a splitting of the ^{13}C line, which for a single crystal gives two peaks whose position and separation depend on the orientation of the crystal in the magnetic field. For powders, all orientations are present, and the observed pattern is a broad envelope of overlapping peaks. Mathematically, the dipole−dipole interaction has the form

$$\text{dipole−dipole interaction} \propto \frac{\gamma_I \gamma_S \hbar}{r^3} \left[1 - 3 \cos^2\theta\right] \times \left[\text{spin terms}\right] \quad (8.4)$$

The variables in Eq. (8.4) are defined in Fig. 8.3. Equation (8.4) and Fig. 8.3 show that the magnitude of the static dipole−dipole

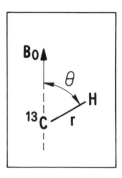

Fig. 8.3. Representation of an internuclear C—H vector of length r, making an angle θ with the magnetic field.

interaction is inversely proportional to the cube of the internuclear distance, and is related to the angle that the C—H internuclear vector makes with the magnetic field direction. In solids the orientations of the C—H internuclear vectors are fixed in the crystallites, and the $(1 - 3 \cos^2\theta)$ term of Eq. (8.4) does not average to zero as it does for solutions. The static dipole—dipole interaction in solids creates a large line broadening, which must be removed in order to obtain high resolution NMR spectra of solids.

8.4 HIGH POWER PROTON DECOUPLING

Early attempts to remove the static dipolar coupling involved modulation of the spin part of Eq. (8.4) through various pulse sequences and modulation of the space term by mechanically spinning the sample about an axis forming an angle with the magnetic field in such a manner that $(1 - 3 \cos^2\theta)$ becomes zero. These methods have largely been supplanted by high power proton decoupling, or *dipolar decoupling*.

Dipolar decoupling is accomplished in a manner analogous to the removal of the scalar, or J-coupling (see Section 2.3.3) in solution state NMR. An rf field is applied to the solid sample at the proton Larmor frequency, as is the case for removing the scalar J-coupling in solutions. For solids, however, the resonant decoupling field must be strong compared to the static dipole—dipole interaction. Instead of the 1 gauss field used for solutions, an approximately 12 gauss (corresponding to approximately 50 kHz) field is required for solid state ^{13}C NMR spectra. The effect of the decoupling field is to remove the static dipolar coupling, as the resonant proton rf field drives the proton spins at a rate that is fast compared to the *ca.* 50 kHz static dipole—dipole interaction. Solid state NMR spectroscopy requires amplifiers capable of providing sufficient power for dipolar decoupling, and sample probes capable of withstanding these high power levels.

The effect of dipolar decoupling on solid state NMR spectra is illustrated in Fig. 8.4. All three ^{13}C NMR spectra are of bulk poly(butylene) terephthalate. The top spectrum, consisting of lines which barely rise above the baseline, is obtained without dipolar decoupling, whereas in the middle spectrum dipolar decoupling is

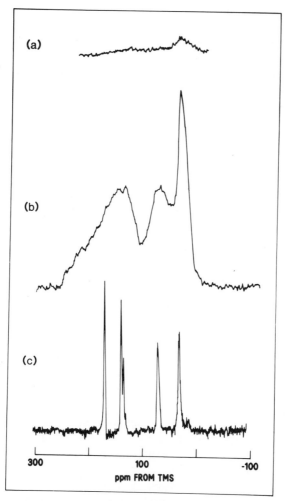

Fig. 8.4. Carbon-13 NMR spectra of bulk poly(butylene terephthalate). (a) Spectrum obtained using low power decoupling. (b) Spectrum obtained using high power (dipolar) decoupling. The primary source of line broadening in this spectrum arises from the chemical shift anisotropy; (c) Spectrum obtained using dipolar decoupling, in addition to magic angle spinning (MAS). MAS reduces the overlapping chemical shift anisotropies to their isotropic values.

employed. [The bottom spectrum combines magic angle spinning (MAS) with dipolar decoupling. MAS is discussed in Section 8.6.] The spectrum in Fig. 8.4b still does not exhibit the high resolution associated with solution state ^{13}C NMR spectra. Although the static dipole–dipole interactions have been removed in this spectrum, the lines are still broad, primarily because of *chemical shift anisotropy*.

8.5 CHEMICAL SHIFT ANISOTROPY

Magnetic shielding and the chemical shift are introduced in Chapter 2. The fact that nuclei in different electronic environments have different isotropic chemical shifts is used repeatedly throughout this book to characterize polymer structure and conformation. The chemical shift arises because the electrons about a particular nucleus interact simultaneously with the nucleus and with the magnetic field, B_0. The external magnetic field produces electric currents in the molecule, and these currents produce a local magnetic field at the nucleus. The *chemical shift tensor* describes the molecular orientation and magnitude of this three-dimensional local field. [Mathematically, a tensor is represented by a set of components that have a definite transformation law under rotations of the coordinate system. The mathematically inclined reader is referred to Brink and Satchler (1962) for further reading on tensors.] Molecules in solution tumble isotropically, averaging the chemical shift tensor to its isotropic value. Thus, the isotropic chemical shift is the only part of the chemical shift tensor observed in solution state NMR. However, in general the chemical shift interaction has the form

$$\text{chemical shift} \propto \gamma \hbar \, \sigma B_0 \tag{8.5}$$

where

$$\sigma = \sigma_{11}\lambda_{11}^2 + \sigma_{22}\lambda_{22}^2 + \sigma_{33}\lambda_{33}^2 \tag{8.6}$$

The σ's are the principal values of the chemical shift tensor and describe the magnitude of the tensor in three mutually perpendicular directions. The λ's are the direction cosines specifying the orientation of the molecular principal axis system with respect to the external field. The chemical shift observed in solution state NMR, or the isotropic chemical shift, is one-third of the sum of the diagonal elements of the tensor or matrix (the sum of the diagonal elements of the tensor is called the *trace*):

$$\sigma_i = (1/3) \, (\sigma_{11} + \sigma_{22} + \sigma_{33}) \tag{8.7}$$

Equations (8.5) and (8.6) indicate that the chemical shift of a particular nucleus in the solid state depends upon the orientation of the nucleus with respect to the magnetic field. This means that the chemical shift of a particular carbon in a single crystal (where all of the carbons have the same orientation with respect to the magnetic field) will change as the crystal is rotated in the field. If one has a powder, rather than a single crystal, all crystallite orientations are present and the resultant NMR spectrum describes the chemical shift tensor powder pattern. Figure 8.5 illustrates two theoretical chemical shift tensor powder patterns. The top one is axially asymmetric and the bottom one is axially symmetric. The principal values σ_{11}, σ_{22}, and σ_{33} are usually expressed in ppm and can be measured directly from the singularities or discontinuities in simple spectra. They are shown in Fig. 8.5. The isotropic chemical shifts $[(1/3)\,(\sigma_{11} + \sigma_{22} + \sigma_{33})]$ are indicated by dotted lines in Fig. 8.5. In the axially symmetric case, the position of $\sigma_{\parallel}$ represents the observed frequency when the principal axis system is parallel with the field direction, and $\sigma_{\perp}$ when the orientation is perpendicular to the field. The effect of molecular motion on the chemical shift tensor powder patterns is to cause a partial narrowing. The exact shape of the motionally narrowed chemical shift tensor contains information concerning the axis and

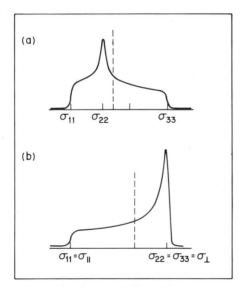

Fig. 8.5. Theoretical broadened chemical shift tensor powder patterns for (a) an axially asymmetric and (b) axially symmetric tensor. The dotted lines represent the isotropic chemical shifts.

angular range of the motion. Clearly, the chemical shift tensor contains considerable angular-dependent structural information. Table 8.1 contains the principal values for some representative chemical shift tensors. In general, carbonyl, carboxyl, and aromatic carbons have the largest anisotropies $(|\sigma_{11} - \sigma_{33}|)$, or powder pattern widths (approximately 180−250 ppm). The anisotropies for methyl, methylene, and methine carbons are usually of the order of 30−60 ppm.

Although the chemical shift powder pattern contains a great deal of information, it contributes a broadening in solid state NMR spectra that often obscures the structural information available from the isotropic chemical shifts. Such is the case in Fig. 8.4b, for example. For this reason, high resolution solid state NMR spectra are often obtained using *magic angle spinning*.

8.6 MAGIC ANGLE SPINNING

Magic angle spinning (MAS) is carried out by mechanically spinning a sample about an axis making the "magic angle" (54.7°) with respect to the magnetic field direction (Fig. 8.6). (The origin of the magic 54.7° angle will be derived below.) MAS is not a new technique. In 1959 Andrew *et al.* (1959) used it to narrow the ^{23}Na line in NaCl, and Lowe (1959) independently used MAS to narrow dipolar-broadened ^{19}F lines in CaF_2 and Teflon. Certain sample spinner

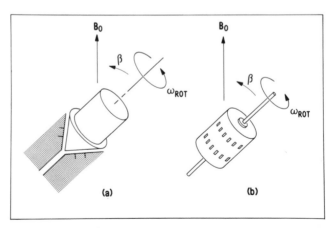

Fig. 8.6. Two designs for magic angle sample spinners. (a) A typical Andrew design sample holder, rotating on air bearings within a stator (shaded part), making an angle β with respect to the external field B_0. The high pressure nitrogen jets are indicated by solid lines on the stator. (b) A Lowe design rotor, supported by phosphor−bronze axles. High pressure air or nitrogen impinges on the flutes.

TABLE 8.1

Magnitudes (ppm) of Some ^{13}C Chemical Shift Tensors[a]

Carbon type	Sample	Temperature (°C)	σ_{11}	σ_{22}	σ_{33}	$|\sigma_{11} - \sigma_{33}|$	Reference
Nonprotonated aromatic	hexamethyl benzene	−186	239	103	31	208	b
	hexamethyl benzene	21	237	71	69	168	b
Protonated aromatic	durene	22	211	135	35	176	c
Carbonyl	acetone	−186	279	265	79	200	d
	carbon monoxide	−269	—	—	—	365	e
	carbon monoxide	−227	—	—	—	335	e
Carboxyl	oxalic acid dihydrate	ambient	249	132	109	140	f
	glycine	ambient	248	180	105	143	g
Methylene	polyethylene	ambient	50	36	12	38	h
	n-eicosane (interior)	−95	50	38	17	33	i
	malonic acid	16	62	51	19	43	j
Methyl	n-eicosane	−95	26	22	3	23	i

[a] All values have been converted to ppm from TMS, using 192.8, 178.4, and 128.7 ppm as the respective chemical shifts for carbon disulfide, acetic acid (carboxyl), and benzene.

[b] Pausak et al. (1974).
[c] Pausak et al. (1973).
[d] Pines et al. (1972b).
[e] Gibson and Scott (1977).
[f] Griffin et al. (1975).
[g] Haberkorn et al. (1981).
[h] Urbino and Waugh (1974).
[i] VanderHart (1976).
[j] Tegenfeldt et al. (1980).

designs bear their names (Fig. 8.6). These early attempts to remove dipolar interactions with MAS were not totally successful as it was impossible to spin the sample rapidly enough to remove the large dipole—dipole interaction. However Schaefer and Stejskal (1976) showed that if one removes the static dipolar interaction by high power proton decoupling (see Section 8.4), the remaining chemical shift anisotropy can be averaged to its isotropic value using MAS.

Under rapid mechanical sample rotation about an angle β with respect to the magnetic field direction (Fig. 8.6) the direction cosines in Eq. (8.6) become time dependent in the rotor period. Taking the time average under rapid sample rotation, Eq. (8.6) becomes

$$\sigma = \tfrac{1}{2} \sin^2\beta \left[\sigma_{11} + \sigma_{22} + \sigma_{33}\right] + \tfrac{1}{2}\left[3 \cos^2\beta - 1\right] \qquad (8.8)$$

$$\times \text{(functions of direction cosines)}$$

The angle β in Eq. (8.8) is the angle that the rotation axis makes with the field direction (Fig. 8.6). When β is 54.7° (the magic angle), $\sin^2\beta$ is 0.66, and the first term in Eq. (8.8) is one-third of the trace of the tensor, or the isotropic chemical shift. The $(3 \cos^2\beta - 1)$ term becomes zero. Using MAS, the chemical shift powder pattern is thus reduced to its isotropic average. The effect of MAS on the ^{13}C NMR spectrum of solid (polybutylene terephthalate) is shown in Fig. 8.4c. In this spectrum the broad overlapping carbonyl and aromatic chemical shift anisotropies of Fig. 8.4b have been reduced to their isotropic averages, and a truly high resolution spectrum is obtained.

8.7 $^1H - ^{13}C$ CROSS POLARIZATION

The spin—lattice relaxation time, or T_1, was introduced in Section 8.2. The T_1 characterizes the return of the spin system to equilibrium after it has been perturbed by a radiofrequency pulse. The spin system must couple to its surroundings, or lattice, in order for relaxation to occur. Molecular motions of megahertz frequencies cause the spin system to couple to its lattice, and efficient relaxation thus results in short T_1 values. The T_1 value for a particular nucleus dictates the rate at which signal averaging can be repeated. The repetition rate becomes an important consideration when observing rare nuclei such as ^{13}C. (The natural abundance of ^{13}C is only 1.1%, and carbon spectra thus require signal averaging.) As expected, solids often have little molecular motion in the megahertz frequency regime, and thus have long carbon T_1 values. [The T_1 for carbon in gem quality diamonds is estimated to be of the order of hours (Hewitt *et al.*, 1982).] Even in solids with methyl groups—which generally rotate rapidly—the T_1 values for the other carbons are prohibitively long. The ^{13}C in the rotating methyl group does not "communicate" its short T_1 value to the other carbons in the sample via spin diffusion, as the nearest ^{13}C neighbor, on the average, is over 0.7 nm distant. In addition, the methyl carbon cannot communicate its short T_1 to its nearby protons, and then from the protons back to other carbons, as the carbon and proton Larmor frequencies are very far apart. Thus there is no overlap in the carbon and proton energies (Fig. 8.7). On the other hand, the *protons* on this methyl group do have short T_1 values, and they can communicate this short T_1 to all the other protons, as the protons are essentially 100% naturally abundant. In general, *proton spin diffusion* causes all of the protons in a solid to have the same T_1 value. In addition, the proton T_1 is generally short compared to most of the carbon T_1 values.

Although there is no energy overlap between ^{13}C and ^{1}H in the laboratory frame (Fig. 8.7), Hartmann and Hahn (1962) demonstrated that an overlap in energies could be made to occur in the *rotating frame*. Thus, energy transfer between nuclei with disparate Larmor frequencies such as protons and ^{13}C could be made to occur when

$$\gamma_C B_{1C} = \gamma_H B_{1H}. \qquad (8.9)$$

Equation (8.9) is called the *Hartmann—Hahn condition*, and results in a match of the rotating frame energies for ^{1}H and ^{13}C. Since γ_H is four times γ_C, the match occurs when the strength of the applied carbon field (B_{1C}) is four times the strength of applied proton field (B_{1H}).

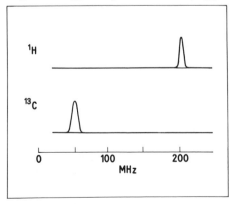

Fig. 8.7. Carbon and proton Larmor frequencies in a 47 kG magnetic field. There is no frequency overlap, and thus no overlap in the energies.

Cross polarization is a method by which polarization is transferred from the abundant spins (^{1}H, in this case) to the rare spins (here, ^{13}C) via Hartmann—Hahn matched energy levels. Because energy is being transferred from the protons to the carbons, the T_1 of the protons dictates the repetition rate for signal averaging.

Cross polarization was introduced by Pines *et al.* in 1972 (Pines *et al.*, 1972a, 1973). The vector diagrams for this double rotating frame experiment are shown in Fig. 8.8, and the pulse sequence is shown in Fig. 8.9. The vector diagrams in Fig. 8.8a show the proton and carbon spin systems equilibrated in the magnetic field. An rf pulse (B_{1H}) is applied at the proton Larmor frequency. It is applied along the x axis and is of sufficient duration to tip the protons 90°, so that they are now along the y axis (Fig. 8.8b). The phase of the proton B_1 field is shifted by 90°, and the protons are then *spin-locked* along the y axis (Fig. 8.8c). [This process is called spin-locking because the proton spins are forced to precess about the y axis of their rotating frame for the duration of the strong B_{1H} pulse. Magnetization decay processes occur during this time, and the interested reader is referred to Schaefer and Stejskal (1976) for details.] The protons precess about

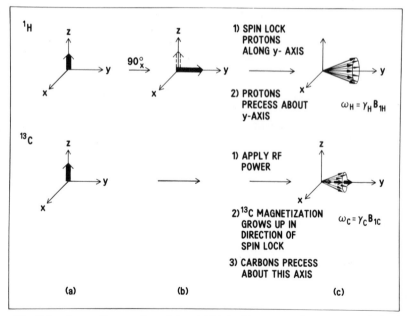

Fig. 8.8.   Vector diagram for a ^{1}H and ^{13}C double rotating frame cross polarization experiment. The carbon frame and the proton frame are rotating at different frequencies. See text for a complete description.

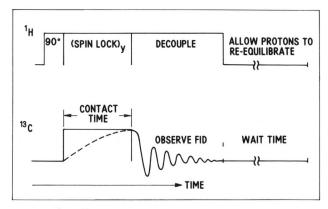

Fig. 8.9.   The ^{1}H—^{13}C cross polarization pulse sequence for solid state NMR.

their rotating frame y axis with frequency $\omega_H = \gamma_H B_{1H}$. Meanwhile, the carbons are put into *contact* with the protons. This is accomplished by turning the carbon B_1 field (B_{1C}) on during the spin-lock time. This causes the ^{13}C magnetization to grow up in the direction of the spin-lock field (see Figs. 8.8c and 8.9). The carbons will precess about this axis in their rotating frame with frequency $\omega_C = \gamma_C B_{1C}$.

How does the proton-to-carbon polarization transfer take place? At this point, the protons are precessing about the B_{1H} field with frequency $\gamma_H B_{1H}$, and the carbons are precessing in the same direction with frequency $\gamma_C B_{1C}$. (See Fig. 8.8c, or in more detail in Fig. 8.10.) When the Hartmann—Hahn condition is matched ($\gamma_H B_{1H} = \gamma_C B_{1C}$) by adjusting the power levels of the B_{1H} and B_{1C} fields, the z-components of both the ^{1}H and ^{13}C magnetizations have the same time dependence (Fig. 8.10). Because the z-component time dependence is common to both spin systems, mutual spin flips can occur between the protons and the carbons. This process can be visualized as a "flow" of polarization from the abundant proton spins to the rare carbon spins.

An alternative way to conceptualize polarization transfer is through a spin temperature or thermodynamic point of view. These concepts are put forth in detail by Pines *et al.* (1973), and the reader is referred there for a fuller discussion.

The advantages of cross polarization are twofold. First, it circumvents the problem of the long carbon T_1 values normally found in solids. The carbons obtain their polarization from the protons, and thus it is the proton T_1 which dictates the cross polarization

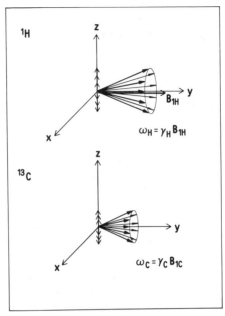

Fig. 8.10. A more detailed representation of Fig. 8.9c. The carbons are precessing with frequency $\omega = \gamma_C B_{1C}$, and the protons with frequency $\omega_H = \gamma_H B_{1H}$. When the Hartmann–Hahn condition is matched ($\gamma_H B_{1H} = \gamma_C B_{1C}$) the two spin systems have z-components with the same frequency dependence. Thus mutual spin flips can occur between ^1H and ^{13}C.

experiment repetition rate. Second, the carbon shows an enhancement in its signal intensity, which can be as large as the ratio of $\gamma_H : \gamma_C$, or a factor of four. Thus, cross polarization results in both a time savings and in an improvement in the signal-to-noise ratio.

8.8 SOLID STATE CARBON 13 NMR OF POLYMERS

Solid state NMR has proved to be of enormous utility in studying polymer conformation, structure, and dynamics. For intractable polymers, for multiphase systems, and for soluble polymers, solid state NMR spectroscopy has provided information otherwise unavailable. For example, Maricq et al. (1978) examined the ^{13}C chemical shifts of cis and trans polyacetylenes, and concluded that these polymers exhibit no metallic character in the undoped state. Others have used solid state ^{13}C NMR to examine polymorphism in a natural polymer, cellulose (Atalla et al., 1980; Earl and VanderHart, 1980). In other applications, ^{13}C NMR of solids has been used as an analytical technique to determine, for example, the protein content in foodstuffs

(Rutar *et al.*, 1980), and the aromatic content in coal (Miknis *et al.*, 1979). We explore four examples of solid state polymer NMR in more detail below.

8.8.1   Carbon 13 NMR of Poly(phenylene oxide) and Other Polymers with Aromatic Residues

Poly(phenylene oxide) provided one of the first examples of NMR observation of conformational nonequivalence in the solid state (Schaefer and Stejskal, 1979). With cross polarization and dipolar decoupling (but without MAS), the ^{13}C NMR spectrum of poly(phenylene oxide) in the solid state exhibits a powder pattern consisting of overlapping chemical shift tensor patterns (Fig. 8.11a). With MAS, the broad pattern is resolved into *six* distinct lines (Fig. 8.11b), rather than the five lines observed in the solution state ^{13}C NMR spectrum (Fig. 8.11c). The line that is doubled has been assigned unambiguously to the protonated aromatic carbons. The carbon−oxygen−carbon bond forms an approximately 109° angle, and in the absence of rapid phenylene ring flips, the protonated aromatic carbons reside in environments different enough to give rise to separate isotropic chemical shifts.

More recently, Garroway and co-workers (1982) studied phenylene ring flips in epoxy resins. They observed "doubled" phenylene resonances for certain epoxy resins at low temperatures. As the temperature was raised, the doubled peaks "coalesced" into one central line, reminiscent of the NMR coalescence phenomena observed in solution state NMR. From their data, Garroway and co-workers were able to estimate activation energies for phenylene ring flips in epoxy resins.

VanderHart *et al.* (1981) have also used variable temperature solid state ^{13}C NMR spectroscopy to study phenylene ring flips in polymers. They find that the phenylene rings of oriented poly(ethylene) terephthalate undergo 180° ring flips at elevated temperatures. These ring flips become fast on the NMR time scale at approximately 77°, a temperature well below the melt temperature. They also find that the single phenylene resonance observed at high temperatures broadens and splits into two peaks as the temperature is lowered. VanderHart's ^{13}C study confirms previous solid state ^{2}H NMR studies on poly(ethylene) terephthalate (Vega, 1980), in which the aromatic residues in the amorphous regions of poly(ethylene) terephthalate were found to undergo 180° phenylene ring flips at elevated temperatures.

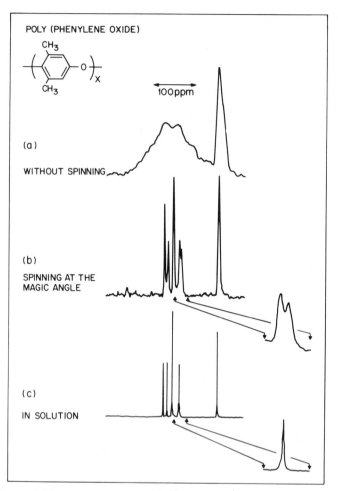

Fig. 8.11. Solid state and solution state ^{13}C NMR spectra of poly(phenylene oxide). (a) Solid state spectrum obtained with cross polarization and dipolar decoupling (but no MAS); (b) spectrum of sample in (a) with MAS; (c) solution state ^{13}C NMR spectrum. [Figure reproduced with modification from Schaefer and Stejskal (1979), with permission of the copyright holder.]

At this time the conformational and free volume factors that enhance or prohibit phenyl and phenylene ring flips are not well understood. As a striking example, at ambient temperatures the phenyl ring in the zwitterion form of phenylalanine is found to undergo 180° ring flips on the NMR time scale (Gall *et al.*, 1981). On the other hand, the aromatic rings in the hydrochloride salt of phenylalanine do not (S. Opella, private communication, 1982).

8.8.2 Isotactic and Syndiotactic Polypropylene

Isotactic polypropylene adopts a 3_1 helical conformation in the solid state, as shown in Fig. 8.12a. The 3_1 helix arises from alternating gauche and trans conformations, and three monomer units are required to form one repeat of the helix (see Section 7.6). Looking along the helix axis, the methyl groups all occupy positions on the surface of the coil. The methyl groups all make the same angle with respect to the helix axis, and thus impart a sense of direction to the helix. The methylene and methine groups are packed, one over the

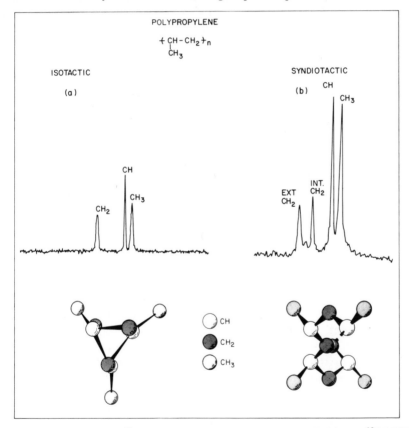

Fig. 8.12.   Solid state ^{13}C NMR spectra of polypropylene. (a) Solid state ^{13}C NMR spectrum of isotactic polypropylene [reproduced with modification from Fleming et al. (1980), by permission of the American Chemical Society] plus representation of the helical conformation of isotactic polypropylene. (b) Solid state ^{13}C NMR spectrum of syndiotactic polypropylene, plus representation of the conformation of syndiotactic polypropylene. [Reproduced from Bunn et al. (1981) by permission of the Royal Chemical Society.]

other, in alternating fashion. In this structure there are three distinct types of carbons, and these give the three resonances observed in Fig. 8.12a (Fleming *et al.*, 1980).

Syndiotactic polypropylene, on the other hand, adopts the conformation shown in Fig. 8.12b. This helix is formed of units with a (*gg*) (*tt*) $\cdots$ conformation, with four monomer units per repeat. In this conformation there are two distinct environments for the methylene groups, external and internal. Accordingly, the ^{13}C NMR spectrum of syndiotactic polypropylene in the solid state exhibits separate resonances for the internal and external methylene groups (Bunn *et al.*, 1981) (Fig. 8.12b). The assignments of the resonances for the internal and external methylene groups have been made based on the γ-gauche shielding effect (Bunn *et al.*, 1981) (see Section 7.8). In this interpretation, the external CH_2 "sees" two trans carbons in the γ-positions, whereas the internal CH_2 sees two gauche carbons in the γ-positions. The γ-gauche shielding effect for polypropylene in solution is 4 ppm (Schilling and Tonelli, 1980), meaning that each γ-gauche interaction is predicted to shift the all-trans position of a polypropylene CH_2 carbon upfield by 4 ppm. According to this interpretation, the peak for the internal CH_2 of syndiotactic polypropylene is predicted to occur 8 ppm upfield (the sum of two γ-gauche effects) from the position of the external CH_2. In the spectrum in Fig. 8.12b, the resonances for the internal and external CH_2's are actually separated by 8.7 ppm.

8.8.3 Crystalline and Noncrystalline Polyethylene

Earl and VanderHart (1979) have examined a variety of polyethylene samples by solid state ^{13}C NMR. In comparing low density branched polyethylene and high density linear polyethylenes with different thermal and pressure histories, they find that the crystalline and noncrystalline regions of the polymer give resonances that are separated by as much as *ca.* 2.4 ppm (see Fig. 8.13). The crystalline regions of the polymer are composed of all-trans sequences, whereas the noncrystalline regions are assumed to contain equilibrium amounts of gauche and trans bonds. The chemical shift differences between the crystalline and noncrystalline regions of polyethylene can be understood in terms of the γ-gauche effect (see Section 7.8).

Solid state ^{13}C NMR cannot be used conveniently to assess the relative amounts of crystalline and noncrystalline components in polyethylene, as the cross polarization technique under-represents the

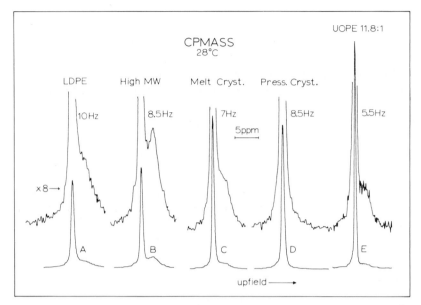

Fig. 8.13.   Solid state ^{13}C NMR spectra of five polyethylene samples. The spectra were obtained with cross polarization, dipolar decoupling, and magic angle spinning. (a) Low density polyethylene; (b−e) high density linear polyethylene [(b) machined from a block of polyethylene molded under mild pressure; (c) melt crystallized and air cooled; (d) pressure crystallized, crystallinity *ca*. 95%; (e) ultraoriented polyethylene]. [Reprinted with permission from Earl and VanderHart (1979), Copyright 1979, American Chemical Society.]

signal arising from the noncrystalline component (Earl and VanderHart, 1979). The underlying reason for the under-representation of the noncrystalline component is related to the mobility of this region.

There also are linewidth differences between the crystalline and noncrystalline components in polyethylene, with the noncrystalline region exhibiting a larger linewidth (Fig. 8.13). The larger linewidth of the noncrystalline component line is attributed both to motional broadening (from motions with frequencies near B_{1H}) and from dispersions in the isotropic chemical shifts of various carbons in the noncrystalline region. Resolution of many of these questions awaits the development of methods and techniques to perform low temperature, magic angle spinning NMR experiments on solid polymers.

8.8.4 Segmented Copolymers

Up until now we have discussed solid state ^{13}C NMR from the point of view of polymer *structure*. It is possible to use NMR to obtain information concerning the *motional dynamics* of solid polymers, as well. Although NMR relaxation in the solid state is outside the scope of this chapter, we shall present an example to illustrate the power of the ^{13}C NMR technique in discerning the local motions of each chemically different carbon in a polymer. [The interested reader is referred to Schaefer and Stejskal (1979) for a clear discussion of NMR relaxation in the solid state.]

Hytrel is a copolyester composed of blocks of "hard" poly(butylene terephthalate) segments and poly(tetramethyleneoxy) terephthalate "soft" segments (Fig. 8.14). The hard and soft segments form separate phases which are thought to consist of interpenetrating crystalline and amorphous domains (Cella, 1977).

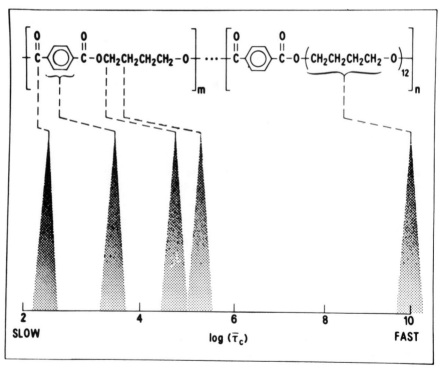

Fig. 8.14. The structure of Hytrel segmented copolymer. The relative amounts of hard (m) and soft (n) segments can vary. Beneath the structure is a schematic representation of the motional rates for each carbon.

In addition to the Hytrel structure, Fig. 8.14 also shows a schematic representation of the rates of motions of the various carbons. These rates were estimated from solid state ^{13}C NMR data, and we shall summarize here the manner in which this was done.

It was found that the soft segment carbons of the solid block copolymer give sharp ^{13}C NMR signals *without* high power decoupling and *without* cross polarization (see Section 8.3) (Jelinski et al., 1981). The intensity of these signals is proportional to the mole fraction of soft segment in the copolymer (Fig. 8.15). This result means that the soft segment carbons have sufficient mobility to significantly average out the strong (approximately 50 kHz) $^1H-^{13}C$ static dipole—dipole interaction. Similar experiments on the homopolymer poly(butylene terephthalate) show very little signal intensity in this region (Fig. 8.15e). Relaxation studies (T_1 measurements and nuclear

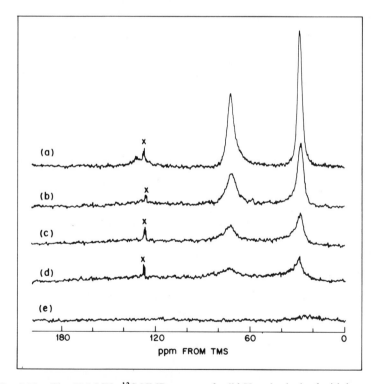

Fig. 8.15.   The 50.3 MHz ^{13}C NMR spectra of solid Hytrel, obtained with low power decoupling and no MAS. (a) 44 wt% hard segments; (b) 57 wt% hard; (c) 75 wt% hard, (d) 81 wt% hard; (e) poly(butylene) terephthalate. The peaks marked with X arise from an internal capillary containing benzene-d_6.

Overhauser experiments) at variable temperatures and variable magnetic field strengths show that these carbons are involved in motions characterized by a distribution of motional frequences, with a mean correlation time of 5×10^{-11} sec at 34°C (Jelinski et al., 1982a). (The *correlation time* can be conceptualized as the amount of time the carbon can be considered stationary, before it moves on to another position.) Further, the *rate* of motions, as determined by the T_1, does not change with variations in the amount of hard segment in the polymer. However, the linewidths for the soft segment carbons are a linear function of the average hard block length of the polymer, suggesting that the *angular range* of these motions becomes increasingly restricted as the hard segment content of the polymer increases (Jelinski et al., 1982b).

Chemical shift anisotropy considerations and relaxation studies were used to characterize molecular motions of the hard segment carbons. For the poly(butylene) terephthalate homopolymer, the widths of the CH_2 carbon lines, without MAS, were found to be significantly different. At ambient temperatures, OCH_2 linewidth, or chemical shift anisotropy, was only slightly reduced from values normally expected for static OCH_2 carbons. However, the $-CH_2CH_2CH_2-$ carbon linewidth was two to threefold narrower than expected. (See Table 8.1 for representative chemical shift anisotropies.) This result suggests that the central CH_2 carbons are involved in motions that do not affect the OCH_2 carbons to a large extent (Jelinski, 1981). Relaxation studies on the poly(butylene terephthalate) homopolymer confirm this finding, as the central CH_2 carbons have a T_1 that is half as large as the T_1 for the OCH_2 carbons. Relaxation studies on the segmented copolymer suggest that these motions occur, and may even be enhanced, in the segmented copolymer (Jelinski and Dumais, 1981). There is also NMR relaxation evidence that these motions and motional differences persist in solution (Jelinski et al., 1982a).

Chemical shift anisotropy measurements were also used to deduce the motional dynamics of the carbonyl and aromatic carbons. These types of carbons generally have large chemical shift anisotropies (see Table 8.1), and these anisotropies are sensitive to low frequency (10's of kHz) motions. Although the methods for reconstruction of the chemical shift anisotropies are outside the scope of this chapter, the results are shown graphically in Fig. 8.16. Here, the component chemical shift anisotropies for the carbonyl, aromatic, and non-protonated aromatic carbons are shown individually (Fig. 8.16b).

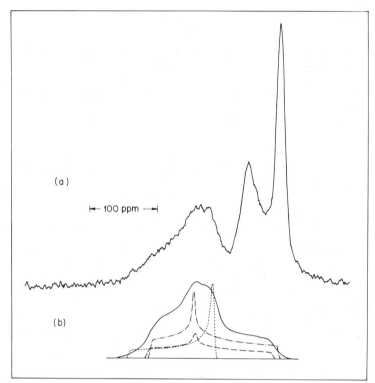

Fig. 8.16.   Top: Experimental ^{13}C NMR spectrum of a 81 wt% hard sample of Hytrel. This spectrum was obtained with cross polarization and high power decoupling, but without MAS. Bottom: calculated carbonyl (---), protonated aromatic ($\cdot-\cdot$), and nonprotonated aromatic (— — —) chemical shift anisotropies for the 81 wt% hard sample.

When summed together, they match very well with the carbonyl and aromatic portion of the experimental NMR spectrum (Fig. 8.16a). The individual chemical shift anisotropies ($|\sigma_{11}-\sigma_{33}|$) were interpreted by comparing them to model compound anisotropies. Although the carbonyl carbon was found to be axially symmetric (Figs. 8.5 and 8.16), its width is similar to static rigid lattice chemical shift anisotropies, ruling out the possibility of its involvement in large-scale "crank-shaft" type motions. The protonated aromatic carbons, on the other hand, show a partially collapsed chemical shift anisotropy, suggesting that there is motion of the phenyl groups about the 1,4-phenylene axis (Jelinski, 1981; Jelinski et al., 1982b).

Summing together the above data, one is able to compile a schematic representation of the segmented copolymer motions, shown in Fig. 8.14.

REFERENCES

Andrew, E. R., Bradbury, A., and Eades, R. G. (1959). *Nature (London)* **183**, 1802.

Atalla, R. H., Gast, J. C., Sindorf, D. W., Bartuska, V. J., and Maciel, G. E. (1980). *J. Am. Chem. Soc.* **102**, 3249.

Brink, D. M., and Satchler, G. R. (1962). "Angular Momentum." Clarendon Press, Oxford.

Bunn, A., Cudby, E. A., Harris, R. K., Packer, K. J., and Say, B. J. (1981). *Chem. Comm.* 15.

Cella, R. J. (1977). *Encycl. Polymer Sci. Technol. Suppl.* **2**, 485.

Cochran, W. T., Cooley, J. W., Faven, D. L., Helms, H. D., Kaenel, R. A., Lang, W. W., Maling, G. C., Jr., Nelson, D. E., Rader, C. M., and Welch, P. D. (1967). *Proc. IEEE* **55**, 1664.

Cooley, J. W., and Tukey, J. W. (1965). *Math. Comput.* **19**, 297.

Earl, W. L., and VanderHart, D. L. (1979). *Macromolecules* **12**, 762.

Earl, W. L., and VanderHart, D. L. (1980). *J. Am. Chem. Soc.* **102**, 3251.

Fleming, W. W., Fyfe, C. A., Kendrick, R. D., Lyerla, J. R., Jr., Vanni, H., and Yannoni, C. S. (1980). *In* "Polymer Characterization by ESR and NMR" (A. E. Woodward and F. A. Bovey, eds.), p. 212. ACS Symposium Series 142.

Gall, C. M., Diverdi, J. A., and Opella, S. J. (1981). *J. Am. Chem. Soc.* **103**, 5039.

Garroway, A. N., Ritchey, W. M., and Moniz, W. B. (1982). *Macromolecules*, in press.

Gibson, A. A., and Scott, T. A. (1977). *J. Mag. Res.* **27**, 29.

Griffin, R. G., Pines, A., Pausak, S., and Waugh, J. S. (1975). *J. Chem. Phys.* **63**, 1267.

Haberkorn, R. A., Stark, R. E., Van Willigen, H., and Griffin, R. G. (1981). *J. Am. Chem. Soc.* **103**, 2534.

Hartmann, S. R., and Hahn, E. L. (1962). *Phys. Rev.* **128**, 2042.

Hewitt, J. M., Henrichs, P. M., Young, R. H., and Cofield, M. L. (1982). *Proc. 23rd Experimental NMR Conference, Madison*, Wisconsin, Abstract A-26.

Jelinski, L. W. (1981). *Macromolecules* **14**, 1341.

Jelinski, L. W., and Dumais, J. J. (1981). *Polymer Preprints* **22**(2), 273.

Jelinski, L. W., Schilling, F. C., and Bovey, F. A. (1981). *Macromolecules*, **14**, 581.

Jelinski, L. W., Dumais, J. J., Watnick, P. I., Engel A., and Sefcik, M.D. (1982a). *Macromolecules*, in press.

Jelinski, L. W., Dumais, J. J., and Engel, A. (1982b). *Macromolecules*, in press.

Lowe, I. J. (1959). *Phys. Rev. Lett.* **2**, 285.

Maricq, M. M., Waugh, J. S., MacDiarmid, A. G., Shirakawa, H., and Haeger, A. J. (1978). *J. Am. Chem. Soc.* **100**, 7729.

Miknis, F. P., Bartuska, V. J., and Maciel, G. E. (1979). *Am. Lab.* November 1979.

Pausak, S., Pines, A., and Waugh, J. S. (1973). *J. Chem. Phys.* **59**, 591.

Pausak, S., Tegenfeldt, J., and Waugh, J. S. (1974). *J. Chem. Phys.* **61**, 1338.

Pines, A., Gibby, M. G., and Waugh, J. S. (1972a). *J. Chem. Phys.* **56**, 1776.

Pines, A., Gibby, M. G., and Waugh, J. S. (1972b). *Chem. Phys. Lett.* **15**, 373.

Pines, A., Gibby, M. G., and Waugh, J. S. (1973). *J. Chem. Phys.* **59**, 569.

Rutar, V., Blinc, R., and Ehrenberg, L. (1980). *J. Mag. Res.* **40**, 225.

Schaefer, J., and Stejskal, E. O. (1976). *J. Am. Chem. Soc.* **98**, 1031.

Schaefer, J., and Stejskal, E. O. (1979). *In* "Topics in Carbon-13 NMR Spectroscopy" (G. C. Levy, ed.), Vol. 3, pp. 283-324. Wiley, New York.

Schilling, F. C., and Tonelli, A. E. (1980). *Macromolecules* **13**, 270.

Tegenfeldt, J., Feucht, H., Ruschitzka, G., and Haeberlen, U. (1980). *J. Mag. Res.* **39**, 509.

Urbino, J., and Waugh, J. S. (1974). *Proc. Nat. Acad. Sci. USA* **71**, 5062.
VanderHart, D. L. (1976). *J. Chem. Phys.* **64**, 830.
VanderHart, D. L., Böhm, G. G. A., and Mochel, V. D. (1981). *Polymer Preprints* **22** (2), 261.
Vega, A. J. (1980). *Proc. 21st Experimental NMR Conference, Tallahassee*, Florida.

INDEX

X

Z